T0214606

Communications
in Computer and Information Science   833

Commenced Publication in 2007
Founding and Former Series Editors:
Phoebe Chen, Alfredo Cuzzocrea, Xiaoyong Du, Orhun Kara, Ting Liu,
Dominik Ślęzak, and Xiaokang Yang

More information about this series at http://www.springer.com/series/7899

Alvaro David Orjuela-Cañón
Juan Carlos Figueroa-García
Julián David Arias-Londoño (Eds.)

Applications of Computational Intelligence

First IEEE Colombian Conference, ColCACI 2018
Medellín, Colombia, May 16–18, 2018
Revised Selected Papers

 Springer

Editors
Alvaro David Orjuela-Cañón
Universidad Antonio Nariño
Bogotá, Colombia

Julián David Arias-Londoño
Universidad de Antioquia
Medellín, Colombia

Juan Carlos Figueroa-García ⓘ
Universidad Distrital Francisco
José de Caldas
Bogotá, Colombia

ISSN 1865-0929   ISSN 1865-0937 (electronic)
Communications in Computer and Information Science
ISBN 978-3-030-03022-3   ISBN 978-3-030-03023-0 (eBook)
https://doi.org/10.1007/978-3-030-03023-0

Library of Congress Control Number: 2018959140

Preface

The use of computational intelligence (CI) techniques in solving real-world and engineering problems is not alien to the Latin America (LA) region. Many LA scientists have focused their efforts on the CI field as a way to deal with problems of interest for the international community but also of great impact in the LA region. Many different areas including optimization of energy and transportation systems, computer-aided medical diagnoses, bioinformatics, mining of massive data sets, robotics and automatic surveillance systems, among many others, are commonly addressed problems from this part of the world, because of the great potential these applications could also have in developing countries. An increasing number of PhD graduates and PhD programs/students in LA related to CI and computer sciences are driving research in CI to increasingly higher standards and making it a common engineering technique in LA.

This way, the First IEEE Colombian Conference on Computational Intelligence (ColCACI 2018) offered a space to all scientists working on applications/theory of CI techniques. Although ColCACI started as a Colombian conference, we received 60 papers by authors from 12 countries, making ColCACI an international forum for CI researchers and practitioners to share their more recent advancements and results. This proceedings volume includes the 17 best papers presented at the conference.

We received a positive feedback from all participants, and so we will keep working on offering ColCACI in future editions and aim to grow. Finally, we would like to thank the IEEE Colombia Section, the Universidad de Antioquia (UdeA), the Universidad Antonio Nariño (UAN), the Universidad Distrital Francisco José de Caldas (UDistrital), all volunteers, all participants, and the whole crew that worked together to host a successful conference. See you at ColCACI 2019!

May 2018

Alvaro David Orjuela-Cañón
Julián David Arias-Londoño
Juan Carlos Figueroa-García

Organization

General Chair

Alvaro David Orjuela-Cañón — Universidad Antonio Nariño, Colombia

Technical Co-chairs

Julián David Arias-Londoño — Universidad de Antioquia, Colombia

Juan Carlos Figueroa-García — Universidad Distrital Francisco José de Caldas, Colombia

Keynote and Tutorials Chair

Fabián Peña — Universidad de los Andes, Colombia

Publication Chairs

Diana Briceño — Universidad Distrital Francisco José de Caldas, Colombia

Alvaro David Orjuela-Cañón — Universidad Antonio Nariño, Colombia

Financial Chair

José David Cely — Universidad Distrital Francisco José de Caldas, Colombia

Webmaster

Fabian Martinez — IEEE Colombia, Colombia

Diego Ruiz — Universidad Central, Colombia

Program Committee

Alvaro David Orjuela Cañón — Universidad Antonio Nariño, Colombia

Julián David Arias Londoño — Universidad de Antioquia, Colombia

Juan Carlos Figueroa Garcia — Universidad Distrital Francisco José de Caldas, Colombia

Jose Manoel de Seixas — Universidade Federal de Rio de Janeiro, Brazil

Danton Ferreira — Universidade Federal de Lavras, Brazil

Efren Gorrostieta | Universidad Autónoma de Queretaro, Mexico
Cristian Rodríguez Rivero | UCDavis Center for Neuroscience, USA
Jose Alfredo Costa | Universidade Federal do Rio Grande do Norte, Brazil
Javier Mauricio Antelis | Instituto Tecnológico de Monterrey, Mexico
Leonardo Forero Mendoza | Pontificia Universidade Católica do Rio de Janeiro, Brazil
Carmelo Bastos Filho | Universidade de Pernambuco, Brazil
Edgar Sánchez | CINVESTAV, Unidad Guadalajara, Mexico
Guilherme Alencar Barreto | Universidade Federal do Ceará, Brazil
Gonzalo Acuña Leiva | Universidad de Santiago de Chile, Chile
Millaray Curilem | Universidad de la Frontera, Chile
Carlos Alberto Cobos Lozada | Universidad del Cauca, Colombia
Juan Bernardo Gómez Mendoza | Universidad Nacional de Colombia - Sede Manizales, Colombia
Diego Peluffo Ordóñez | Universidad Técnica del Norte, Ecuador
Gerardo Muñoz Quiñones | Universidad Distrital Francisco José de Caldas, Colombia
Jorge Eliécer Camargo Mendoza | Universidad Antonio Nariño, Colombia
Claudia Victoria Isaza Narvaez | Universidad de Antioquia, Colombia
Sandra Esperanza Nope Rodríguez | Universidad del Valle, Colombia
Juan Antonio Contreras Montes | Universidad Tecnológica de Bolivar, Colombia
Jesús Alfonso López Sotelo | Universidad Autónoma de Occidente, Colombia
Cesar Hernando Valencia Niño | Universidad Santo Tomás - Sede Bucaramanga, Colombia
Miguel Melgarejo Rey | Universidad Distrital Francisco José de Caldas, Colombia
Wilfredo Alfonso Morales | Universidad del Valle, Colombia
Alfonso Perez Gama | Fundación Educación Superior San Jose, Colombia
Néstor Darío Duque Méndez | Universidad Nacional de Colombia - Sede Manizales, Colombia
Mauricio Orozco Alzate | Universidad Nacional de Colombia - Sede Manizales, Colombia
César Germán Castellanos Domínguez | Universidad Nacional de Colombia - Sede Manizales, Colombia
Víctor Hugo Grisales Palacio | Universidad Nacional de Colombia - Sede Bogotá, Colombia
Genaro Daza Santacoloma | Instituto de Epilepsia y Parkinson del Eje Cafetero S.A. - Pereira, Colombia
Fabio A. González | Universidad Nacional de Colombia - Sede Bogotá, Colombia
Fernando Lozano | Universidad de Los Andes, Colombia

Pablo Andrés Arbelaez Escalante	Universidad de Los Andes, Colombia
Humberto Loaiza	Universidad del Valle, Colombia
Eduardo Francisco Caicedo Bravo	Universidad del Valle, Colombia
Juan Carlos Niebles	Universidad del Norte, Colombia
Carlos Andrés Quintero Peña	Universidad Santo Tomás - Sede Bogotá, Colombia
Alexander Molina Cabrera	Universidad Tecnológica de Pereira, Colombia
Luiz Pereira Caloba	Universidade Federal de Rio de Janeiro, Brazil
Leonardo Forero Mendoza	Universidade Estadual de Rio de Janeiro, Brazil
Alvaro Gustavo Talavera	Universidad del Pacífico, Peru
Efraín Mayhua-López	Universidad Católica San Pablo, Peru
Yván Tupac	Universidad Católica San Pablo, Peru
Ana Teresa Tapia	Escuela Superior Politécnica del Litoral, Ecuador
Miguel Núñez del Prado	Universidad del Pacífico, Peru
Heitor Silvério Lopes	Universidade Tecnológica Federal de Paraná, Brazil
Waldimar Amaya	ICFO-The Institute of Photonic Sciences, Spain
Leonardo Franco	Universidad de Málaga, Spain
Carlos Andrés Peña	University of Applied Sciences Western Switzerland, Switzerland
Edwin Alexander Cerquera	University of Florida, USA
Nadia Nedjah	Universidade Estadual do Río de Janeiro, Brazil
María Daniela López de Luise	CI2S Lab, Argentina
Gustavo Eduardo Juarez	Universidad Nacional de Tucuman, Argentina
Ernesto Cuadros	Universidad Católica San Pablo, Perú

Contents

Artificial Neural Networks

Computational Intelligence

Computer Science

Artificial Neural Networks

Spatial and Temporal Feature Extraction Using a Restricted Boltzmann Machine Model

Jefferson Hernandez and Andres G. Abad[(✉)]

Escuela Superior Politecnica del Litoral (ESPOL), Campus Gustavo Galindo Velasco, 09-01-5863, Guayaquil, Ecuador
{jefehern,agabad}@espol.edu.ec

Abstract. A restricted Boltzmann machine (RBM) is a generative neural-network model with many applications, such as, collaborative filtering, acoustic modeling, and topic modeling. An RBM lacks the capacity to retain memory, making it inappropriate for dynamic data modeling as in time-series or video analysis. In this work we address this issue by proposing the p-RBM model: a generalization of the regular RBM model capable of retaining memory of p past states. We further show how to train the p-RBM model using contrastive divergence and test our model on the problem of recognizing human actions from video data using unsupervised feature extraction. Obtained results show that the p-RBM offers promising capabilities for feature-learning in classification applications.

Keywords: Restricted Boltzmann machines · Neural networks
Sequential data · Video · Human action recognition
Unsupervised feature extraction

1 Introduction

A restricted Boltzmann machine (RBM) is a generative neural-network model used to represent the distribution of random observations and to extract features of those observations in an unsupervised setting. In RBM, the independence structure between its variables is modeled through the use of latent variables (see Fig. 1). Restricted Boltzmann machines were introduced in [17]; although it was not until Hinton proposed contrastive divergence (CD) as a training technique in [9], that their true potential was unveiled.

Restricted Boltzmann machines have proven powerful enough to be effective in diverse settings. Applications of RBM include: collaborative filtering [14], acoustic modeling [4], human-motion modeling [19, 20], and music generation [2].

Restricted Boltzmann machine variations have gained popularity over the past few years. Two variations relevant to this work are: the RNN-RBM [4] and the conditional RBM [19]. The RNN-RBM estimates the density of multivariate time series data by pre-training an RBM to extract features and then training a

© Springer Nature Switzerland AG 2018
A. D. Orjuela-Cañón et al. (Eds.): ColCACI 2018, CCIS 833, pp. 3–13, 2018.
https://doi.org/10.1007/978-3-030-03023-0_1

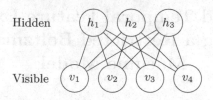

Fig. 1. Example of an RBM with 4 visible units and 3 hidden units.

recurrent neural network (RNN) to make predictions; this allows the parameters of the RBM to be kept constant serving as a prior for the data distribution while the biases are allowed to be modified by the RNN to convey temporal information. Conditional RBMs, on the other hand, estimate the density of multivariate time series data by connecting past and present units to hidden variables. However, this makes the conditional RBM unsuitable for traditional CD training [12].

In this work, we focus on the use of a modified RBM model that does not keep its parameters constant in each time step (unlike the RNN-RBM); and that adds hidden units for past interactions and lacks connections between past and future visible units (unlike the conditional RBM). Our model is advantageous because of two factors: (1) its structure allows the modeling of temporal and spatial structures within a sequence through the extraction of meaningful features that can be used for prediction and classification tasks; and (2) its topology can be easily changed or learned because it is controlled by a single set of parameters, allowing for many models to be readily tested and compared. We show the performance of our model by applying it to the problem of Human Action Recognition (HAR), i.e. classifying human actions in video sequences. The present work is an extended version of our paper [8].

The rest of this work is organized as follows. Section 2 presents a review of RBMs describing their energy function and training method trough CD. Section 3 introduces our proposed model, called the p-RBM, which can be viewed as an ensemble of RBMs with the property of recalling past interactions. In Sect. 4 we apply our model to recognize human actions in video sequences. Conclusions and future research directions are provided in Sect. 5.

2   The Restricted Boltzmann Machine Model

Restricted Boltzmann machines are formed by n visible units, which we represent as $\mathbf{v} \in \{0, 1\}^n$; and m hidden units, which we represent as $\mathbf{h} \in \{0, 1\}^m$. The joint probability of these units is modeled as

$$p(\mathbf{v}, \mathbf{h}) = \frac{1}{Z} e^{-E(\mathbf{v}, \mathbf{h})};$$ (1)

where the energy function $E(\mathbf{v}, \mathbf{h})$ is given by

$$E(\mathbf{v}, \mathbf{h}) = -\mathbf{a}^\mathsf{T}\mathbf{v} - \mathbf{b}^\mathsf{T}\mathbf{h} - \mathbf{v}^\mathsf{T}\mathbf{W}\mathbf{h};$$ (2)

matrix $\mathbf{W} \in \mathbb{R}^{n \times m}$ represents the interaction between visible and hidden units; $\mathbf{a} \in \mathbb{R}^n$ and $\mathbf{b} \in \mathbb{R}^m$ are the biases for the visible and hidden units, respectively; and Z is the partition function defined by

$$Z(\mathbf{a}, \mathbf{b}, \mathbf{W}) = \sum_{\mathbf{v},\mathbf{h}} e^{-E(\mathbf{v},\mathbf{h})}. \tag{3}$$

The bipartite structure of an RBM is convenient because it implies that the visible units are conditionally independent given the hidden units and vice versa. This ensures that the model is able to capture the statistical dependencies between the visible units, while remaining tractable.

Training an RBM is done via CD, which yields a learning rule equivalent to subtracting two expected values: one with respect to the data and the other with respect to the model. For instance, the update rule for $\mathbf{W}$ is

$$\Delta \mathbf{W} = \langle \mathbf{v}\mathbf{h}^{\mathsf{T}} \rangle_{\text{Data}} - \langle \mathbf{v}\mathbf{h}^{\mathsf{T}} \rangle_{\text{Model}}. \tag{4}$$

The first term in Eq. (4) is the expected value with respect to the data and the second is the expected value with respect to the model. For a detailed introduction to RBMs the reader is referred to [5].

3   The p-RBM Model

We generalized the RBM model by constructing an ensemble of RBMs, each one representing the state of the system at p connected moments in time. One way to visualize this is to think of it as p distinct RBMs connected together so as to model the correlation between moments in time. Each RBM contains a representation of the object of interest at different time steps, for example pixels of a video frame, or the value of a dynamic economic index. We then added connections between all the visible and hidden units across time to model their autocorrelation (see Fig. 2).

Our model resembles a Markov chain of order p, because the RBM at time t is conditionally independent of the past, given the previous p RBMs. We showed that even with these newly added time connections the model remains tractable and can be trained in a similar fashion to that of a single RBM.

For convenience, we bundled the visible and hidden units in block vectors denoted $\tilde{\mathbf{v}}$ and $\tilde{\mathbf{h}}$, respectively; we also included a vector of ones of appropriate size, to account for biases interactions; giving

$$\tilde{\mathbf{v}} = \begin{bmatrix} \mathbf{v}_t \vdots \mathbf{v}_{t-1} \vdots \cdots \vdots \mathbf{v}_{t-p} \vdots \mathbf{1} \end{bmatrix}^{\mathsf{T}}, \text{ and} \tag{5}$$

$$\tilde{\mathbf{h}} = \begin{bmatrix} \mathbf{h}_t \vdots \mathbf{h}_{t-1} \vdots \cdots \vdots \mathbf{h}_{t-p} \vdots \mathbf{1} \end{bmatrix}^{\mathsf{T}}. \tag{6}$$

The parameters can be bundled in a block matrix as follows

$$\tilde{\mathbf{W}} = \begin{bmatrix} \mathbf{W}^{v_t,h_t} & \mathbf{W}^{v_t,h_{t-1}} & \cdots & \mathbf{W}^{v_t,h_{t-p}} & \mathbf{W}^{v_t} \\ \mathbf{W}^{v_{t-1},h_t} & \mathbf{W}^{v_{t-1},h_{t-1}} & \cdots & \mathbf{W}^{v_{t-1},h_{t-p}} & \mathbf{W}^{v_{t-1}} \\ \vdots & \vdots & \vdots & \vdots & \vdots \\ \mathbf{W}^{v_{t-p},h_t} & \mathbf{W}^{v_{t-p},h_{t-1}} & \cdots & \mathbf{W}^{v_{t-p},h_{t-p}} & \mathbf{W}^{v_{t-p}} \\ \mathbf{W}^{h_t} & \mathbf{W}^{h_{t-1}} & \cdots & \mathbf{W}^{v_{t-p}} & 0 \end{bmatrix}. \tag{7}$$

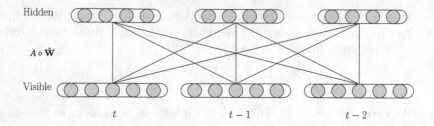

Hidden

$A \circ \hat{\mathbf{W}}$

Visible

t $t-1$ $t-2$

Fig. 2. Example of a p-RBM with $p = 3$ steps in the past.

We added a new hyperparameter $\mathbf{A} \in [0,1]^{p+2,p+2}$ to the model, that acts as a forgetting rate, allowing the model to prioritize connections that are close in time. While this new hyperparameter could be learned through, for instance, cross-validation or bayesian-optimization methods, in this work it is imposed.

We proposed the following structure for matrix $\mathbf{A}$

$$\mathbf{A} = \begin{cases} \alpha^{|i-j|} & i,j \leq p \\ 1 & \text{elsewhere} \end{cases},$$ (8)

for fixed $\alpha \in [0,1]$; or expressed in matrix form as

$$\mathbf{A} = \begin{bmatrix} \alpha^0 & \alpha^1 & \cdots & \alpha^p & 1 \\ \alpha^1 & \alpha^0 & \cdots & \alpha^{p-1} & 1 \\ \vdots & \vdots & \vdots & \vdots & \vdots \\ \alpha^p & \alpha^{p-1} & \cdots & \alpha^0 & 1 \\ 1 & 1 & \cdots & 1 & 1 \end{bmatrix}.$$ (9)

For $\alpha = 1$ the model becomes fully connected, with all connections having the same significance. For $\alpha = 0$ the model is completely disconnected. Thus, matrix $\mathbf{A}$ has some control on the topology of the model.

The energy function for our p-RBM model is given by

$$\begin{aligned} E(\tilde{\mathbf{v}}, \tilde{\mathbf{h}}) &= -\tilde{\mathbf{v}}^{\mathsf{T}}(\mathbf{A} \circ \tilde{\mathbf{W}})\tilde{\mathbf{h}} \\ &= -\sum_{i=0}^{p} \sum_{j=0}^{p} \alpha^{|i-j|} \mathbf{v}_{t-i}^{\mathsf{T}} \mathbf{W}^{v_{t-i},h_{t-j}} \mathbf{h}_{t-j} \\ &\quad - \sum_{i=0}^{p} \mathbf{v}_{t-i}^{\mathsf{T}} \mathbf{W}^{v_{t-i}} - \sum_{j=0}^{p} \mathbf{h}_{t-j}^{\mathsf{T}} \mathbf{W}^{h_{t-j}}, \end{aligned}$$ (10)

where $\circ$ denotes the Hadamard product, indicating element-wise multiplication between matrices.

The energy function given in Eq. (10) considers the effect of the previous interaction. It provides the model with the capacity to include the past, allowing for greater flexibility than that of a typical RBM. As a consequence, the model can, for instance, model high-dimensional time series and non-linear dynamical systems, or reconstruct video from incomplete or corrupted versions.

The joint probability distribution under the model is given by the following Boltzmann-like distribution

$$p(\tilde{\mathbf{v}}, \tilde{\mathbf{h}}) = \frac{1}{Z(\tilde{\mathbf{W}})} e^{-E(\tilde{\mathbf{v}}, \tilde{\mathbf{h}})}, \tag{11}$$

where $E(\tilde{\mathbf{v}}, \tilde{\mathbf{h}})$ is given in Eq. (10), and $Z(\tilde{\mathbf{W}})$ is the partition function defined as

$$Z(\tilde{\mathbf{W}}) = \sum_{\tilde{\mathbf{v}}, \tilde{\mathbf{h}}} e^{-E(\tilde{\mathbf{v}}, \tilde{\mathbf{h}})}. \tag{12}$$

The model structure induces the following relation of conditional independence

$$p(\tilde{\mathbf{h}}|\tilde{\mathbf{v}}) = p(\mathbf{h}_t|\tilde{\mathbf{v}})p(\mathbf{h}_{t-1}|\tilde{\mathbf{v}}) \cdots p(\mathbf{h}_{t-p}|\tilde{\mathbf{v}}). \tag{13}$$

Derivative of the Log-Likelihood of the Model. Given a single training example $\tilde{\mathbf{v}} = \{\mathbf{v}_t, \mathbf{v}_{t-1}, \cdots, \mathbf{v}_{t-p}\}$, the log-likelihood of our model is given by

$$
\begin{aligned}
\ln\mathcal{L}(\tilde{\mathbf{W}}|\tilde{\mathbf{v}}) &= \ln P(\tilde{\mathbf{v}}|\tilde{\mathbf{W}}) \\
&= \ln \frac{1}{Z(\tilde{\mathbf{W}})} \sum_{\tilde{\mathbf{h}}} e^{-E(\tilde{\mathbf{v}}, \tilde{\mathbf{h}})} \\
&= \ln \sum_{\tilde{\mathbf{h}}} e^{-E(\tilde{\mathbf{v}}, \tilde{\mathbf{h}})} - \ln \sum_{\tilde{\mathbf{v}}, \tilde{\mathbf{h}}} e^{-E(\tilde{\mathbf{v}}, \tilde{\mathbf{h}})}.
\end{aligned}
\tag{14}
$$

In what follows, we replace $E(\tilde{\mathbf{v}}, \tilde{\mathbf{h}})$ with E.

Let w be a parameter in $\tilde{\mathbf{W}}$, then the derivative of the log-likelihood w.r.t. w becomes

$$\frac{\partial \ln\mathcal{L}(\tilde{\mathbf{W}}|\tilde{\mathbf{v}})}{\partial w} = -\sum_{\tilde{\mathbf{h}}} p(\tilde{\mathbf{h}}|\tilde{\mathbf{v}}) \frac{\partial E}{\partial w} + \sum_{\tilde{\mathbf{h}}, \tilde{\mathbf{v}}} p(\tilde{\mathbf{h}}, \tilde{\mathbf{v}}) \frac{\partial E}{\partial w}. \tag{15}$$

Equation (15) shows that, as with the RBM, the derivative of the log-likelihood of the p-RBM can be written as the sum of two expectations. The first term is the expectation of the derivative of the energy under the conditional distribution of the hidden variables given an example $\{\mathbf{v}_t, \mathbf{v}_{t-1}, \cdots, \mathbf{v}_{t-p}\}$ from the training set $\mathcal{T}$. The second term is the expectation of the derivative of the energy under the p-RBM distribution [5]. Note that Eq. (15) can be written as

$$\Delta w = -\frac{\partial \ln\mathcal{L}(w|\mathcal{T})}{\partial w} = \left\langle \frac{\partial E}{\partial w} \right\rangle_{\text{Data}} - \left\langle \frac{\partial E}{\partial w} \right\rangle_{\text{Model}}. \tag{16}$$

Equation (16) is the update step corresponding to the p-RBM model and is analogous to Eq. (4).

Contrastive Divergence for the Model. When considering our model, the k-step contrastive divergence (CD_k) becomes

$$CD_k(w, \tilde{\mathbf{v}}^{(0)}) = -\sum_{\tilde{\mathbf{h}}^{(0)}} p(\tilde{\mathbf{h}}^{(0)}|\tilde{\mathbf{v}}^{(0)})\frac{\partial E}{\partial w} + \sum_{\tilde{\mathbf{h}}^{(k)}} p(\tilde{\mathbf{h}}^{(k)}|\tilde{\mathbf{v}}^{(k)})\frac{\partial E}{\partial w}, \tag{17}$$

where a Gibbs chain has been run k times on the second term in order to approximate the expectation of the derivative of the energy under the model, given in Eq. (15).

Training the Model. In order to define a training rule for the model, we propose a way to sample for the CD. The sampling of a block vector of hidden variables can be done from the following block vector of probabilities

$$\mathbb{P}(\tilde{\mathbf{h}}|\tilde{\mathbf{v}}) = \left[p(\mathbf{h}_t|\tilde{\mathbf{v}})\vdots \cdots \vdots p(\mathbf{h}_{t-p}|\tilde{\mathbf{v}})\vdots 1 \right]^{\mathsf{T}}. \tag{18}$$

To sample a vector of visible variables, we construct a block matrix similar to Eq. (7) as

$$\mathbb{P}(\tilde{\mathbf{v}}|\tilde{\mathbf{h}}) = \begin{bmatrix} p(\mathbf{v}_t|\mathbf{h}_t) & \cdots & p(\mathbf{v}_t|\mathbf{h}_{t-p}) & p(\mathbf{v}_t|\tilde{\mathbf{h}}) \\ p(\mathbf{v}_{t-1}|\mathbf{h}_t) & \cdots & p(\mathbf{v}_{t-1}|\mathbf{h}_{t-p}) & p(\mathbf{v}_{t-1}|\tilde{\mathbf{h}}) \\ \vdots & \vdots & \vdots & \vdots \\ p(\mathbf{v}_{t-p}|\mathbf{h}_t) & \cdots & p(\mathbf{v}_{t-p}|\mathbf{h}_{t-p}) & p(\mathbf{v}_{t-p}|\tilde{\mathbf{h}}) \\ 1 & \cdots & 1 & 0 \end{bmatrix}. \tag{19}$$

We may obtain the expected values of the hidden vectors by repeatedly sampling from Eq. (18). For our purposes it is better to place them in a block matrix with the same dimension as $\tilde{\mathbf{W}}$. We call this matrix $\mathbb{E}[\tilde{\mathbf{h}}|\tilde{\mathbf{v}}]$ and it is formed by vertically stacking $p + 2$ block vectors of the form

$$\left[\mathbb{E}[\mathbf{h}_t|\tilde{\mathbf{v}}]\vdots \cdots \vdots \mathbb{E}[\mathbf{h}_{t-p}|\tilde{\mathbf{v}}]\vdots 1 \right]. \tag{20}$$

The derivatives of E can be written as

$$\frac{\partial E}{\partial \tilde{\mathbf{W}}} = \mathbf{A} \circ \tilde{\mathbf{v}} \circ \tilde{\mathbf{h}}. \tag{21}$$

Combining a CD_k, as in Eq. (17), with Eqs. (20) and (21), Eq. (15) becomes

$$\Delta\tilde{\mathbf{W}} = \mathbf{A} \circ (\tilde{\mathbf{v}}^{(0)} \circ \mathbb{E}[\tilde{\mathbf{h}}^{(0)}|\tilde{\mathbf{v}}^{(0)}] - \tilde{\mathbf{v}}^{(k)} \circ \mathbb{E}[\tilde{\mathbf{h}}^{(k)}|\tilde{\mathbf{v}}^{(k)}]). \tag{22}$$

The learning rule is then given by

$$\tilde{\mathbf{W}}^{(\tau+1)} = \tilde{\mathbf{W}}^{(\tau)} + \eta\Delta\tilde{\mathbf{W}}, \tag{23}$$

where $\eta > 0$ is the learning rate.

Note that each item in the block matrix can be learned independently, once the CD step is finalized. This results in a model that is highly suitable for parallelization during training, specially on a GPU where its multicore structure allows the concurrent handling of multiple computational threads.

Algorithm 1 describes the detailed procedure for training the p-RBM model.

Algorithm 1. k-step contrastive divergence for the p-RBM model

---

Input: Training set $\mathcal{T} = \{\mathbf{v}_t, \mathbf{v}_{t-1}, \cdots, \mathbf{v}_{t-p}\}_{t=1}^T$, and a value for α
Output: Updates of all parameters in the model, contained in the block matrix from equation (7)
1: **for all** $\{\mathbf{v}_t, \mathbf{v}_{t-1}, \cdots, \mathbf{v}_{t-p}\} \in \mathcal{T}$ **do**
2:   $\mathbf{v}_t^{(0)}, \mathbf{v}_{t-1}^{(0)}, \cdots, \mathbf{v}_{t-p}^{(0)} \leftarrow \tilde{\mathbf{v}}$
3:     **for** $k = 1, \ldots, K - 1$ **do**
4:       sample $\tilde{\mathbf{h}}^{(k)} \sim \mathbb{P}(\tilde{\mathbf{h}}|\tilde{\mathbf{v}}^{(k-1)})$ as in equation (18)
5:       sample $\tilde{\mathbf{v}}^{(k)} \sim \mathbb{P}(\tilde{\mathbf{v}}|\tilde{\mathbf{h}}^{(k)})$ as in equation (19)
6:     **for** $t = 1, \ldots, n$ **do**
7:       Calculate $\mathbb{E}[\tilde{\mathbf{h}}|\tilde{\mathbf{v}}]$ according to equation (20)
8:       Update $\tilde{\mathbf{W}}$ according to equation (22)
9: **Return** $\tilde{\mathbf{W}}$

---

4   Numerical Results and Analysis

In this section we show the feature-learning capabilities of the p-RBM by applying the model to the problem of unsupervised feature extraction for human action recognition. For an application of the p-RBM model concerning prediction of time-series data the reader is referred to [8].

4.1   Unsupervised Feature Extraction from Video

Human action recognition (HAR) is an active research topic with a large number of applications, such as: video surveillance, video retrieval, assisted living, and human-computer interaction. HAR focuses on the identification of spatio-temporal motion changes using invariant features extracted from 2-D images. Traditional approaches to HAR include the use of hidden Markov models (HMM) on handcrafted features, such as in [3] where a *fast skeletonization* technique was used for feature generation. Other approaches include the use of local spatial and temporal features obtained from detecting image structures within video segments of high variability [11]; they are then combined with support vector machines in [15] to discriminate among six actions.

Unsupervised-learning approaches to HAR have also been proposed in the literature. For instance, in [13] the authors used a Gaussian-RBM to extract spatial features from motion-sensor data to feed an HMM; in [6] a deep Boltzman machine was used in combination with a feed-forward neural network to perform action classification. These techniques, however, are limited in their expressiveness in the sense that they cannot extract time-invariante spatial features. Here, we propose the use of the sequence-modeling capabilities of the p-RBM to extract spatial and temporal features, which is a critical signal to the HAR problem because they provide strong cues for the performed action. For example, a person running will not immediately change the orientation of their legs from one frame to another because actions are correlated between nearby frames.

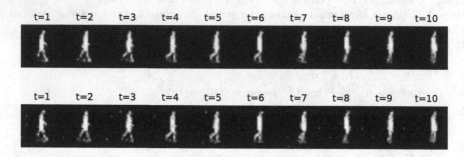

Fig. 3. (Top) Example of a reconstruction of 10 timesteps made by p-RBM on an image in the training dataset. (Bottom) Real sequence of images. Action: *walking*.

We trained our model on the KTH dataset introduced in [15]. This dataset contains six-human actions performed by 25 subjects. For our purposes we turn the problem into a binary classification problem: discriminate between *walking*, and *not walking*. Prior preprocessing included: downsizing the images from a resolution of 160×120 to 28×28 pixels; background subtraction using a simple gaussian-mixture based algorithm; and pixels normalization. Additionally, we created 8,144 sequences for the training set and 4,026 for the testing set. We chose the following set-up for our network: $p = 10$ for the moving window, $n = 784$ for the size of the visible layer, $m = 256$ for the size of the hidden layer, a mixing rate of $k = 1$, and $\alpha = 1$ (i.e., a fully-connected model).

Our model can be used to reconstruct the sequences in the dataset (see Fig. 3, in which we reconstruct a sample from the training set). The features learned by the model are significant as the model is able to reconstruct and de-noise unseen images from new categories (see Fig. 4).

For the classification problem we extracted the features (the $\tilde{h}$ block vector) from all video sequences in the training set and used them to train a multilayer perceptron (MLP) with the architecture described in Table 1. The use of such a simple architecture is motivated by the fact that the features extracted by the p-RBM already summarize the spatial and temporal context of the action performed.

We evaluated the performance of our model by constructing a confusion matrix on the test set, shown in Table 2. The model obtains an overall accuracy of 0.9284.

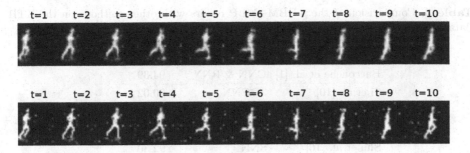

Fig. 4. (Top) Example of a reconstruction of 10 timesteps made by p-RBM on an image outside the training dataset never seen by the p-RBM. (Bottom) Real sequence of images. Action: *running*.

Table 1. Architecture of the MLP that sits on top of the p-RBM used to perform classification. Here BS stands for Batch Size.

Layer	Output shape
Input	(BS, 256, 10)
Flatten	(BS, 2560)
Dense+Relu	(BS, 1024)
Dropout (0.25)	(BS, 1024)
Dense+Relu	(BS, 256)
Dropout (0.25)	(BS, 256)
Dense+Sigmoid	(BS, 1)

Table 2. Confusion matrix predicted action for the KTH dataset in the test set.

Real action	Predicted action	
	Walking	Not walking
Walking	5, 244	104
Not walking	459	5, 244

We present the results of the experiment on the KTH dataset in Table 3, together with other relevant attempted approaches on the same dataset reported in the literature. Table 3 shows that unsupervised featured extraction using the p-RBM is competitive with supervised methods.

Our model surpasses methods that either focus only on temporal features (e.g., [7]); or spatial features (e.g., [16] and [10]). The p-RBM is, however, surpassed by methods that focus on, both, spatial and temporal features, adding more representative power; such as a CNN with assumptions of locality and equivariance [1] and [18]; and an LSTM network with a modified gating scheme that considers the information of salient motion between successive frames [21].

Table 3. Comparision of the p-RBM+MLP results with other methods on the KTH dataset.

Author	Method	Performance
Baccouche et al. [1]	CNN & RNN	94.39
Ji et al. [10]	3DCNN	90.02
Grushin et al. [7]	LSTM	90.70
Veeriah et al. [21]	Differential RNN	93.96
Shu et al. [16]	SNN	92.30
Shi et al. [18]	DTD, DNN	95.60
Ours	**p-RBM+MLP**	**92.84**

5   Conclusion and Future Work

In this work, we proposed an extension to the RBM model, called the p-RBM. We tested our model on the problem of unsupervised feature extraction for human action recognition (HAR) and report positive results. Experimentation with real data showed that our model is suitable for problems involving high-dimensional correlated data by extracting spatio-temporal representations. When compared with other models, the p-RBM showed competitive performance with state-of-the-art results on the HAR task.

The following are possible research directions for our work: (1) generative modeling and distribution learning on the obtained feature space; (2) discovering the most suitable topology for a problem by exploring the space of hyperparameter matrix **A**; and (3) extending our model by considering convolutional variants of the RBM.

References

1. Baccouche, M., Mamalet, F., Wolf, C., Garcia, C., Baskurt, A.: Sequential deep learning for human action recognition. In: Salah, A.A., Lepri, B. (eds.) HBU 2011. LNCS, vol. 7065, pp. 29–39. Springer, Heidelberg (2011). https://doi.org/10.1007/978-3-642-25446-8_4
2. Boulanger-Lewandowski, N., Bengio, Y., Vincent, P.: Modeling temporal dependencies in high-dimensional sequences: application to polyphonic music generation and transcription (2012)
3. Chen, H.S., Chen, H.T., Chen, Y.W., Lee, S.Y.: Human action recognition using star skeleton. In: Proceedings of the 4th ACM International Workshop on Video Surveillance and Sensor Networks, VSSN 2006, pp. 171–178. ACM, New York (2006). https://doi.org/10.1145/1178782.1178808
4. Dahl, G., Ranzato, M.A., Mohamed, A.R., Hinton, G.E.: Phone recognition with the mean-covariance restricted Boltzmann machine. In: Lafferty, J.D., Williams, C.K.I., Shawe-Taylor, J., Zemel, R.S., Culotta, A. (eds.) Advances in Neural Information Processing Systems, vol. 23, pp. 469–477. Curran Associates Inc. (2010)

5. Fischer, A., Igel, C.: Training restricted Boltzmann machines: an introduction. Pattern Recognit. **47**(1), 25–39 (2014)
6. Foggia, P., Saggese, A., Strisciuglio, N., Vento, M.: Exploiting the deep learning paradigm for recognizing human actions. In: 2014 11th IEEE International Conference on Advanced Video and Signal Based Surveillance (AVSS), pp. 93–98. IEEE (2014)
7. Grushin, A., Monner, D.D., Reggia, J.A., Mishra, A.: Robust human action recognition via long short-term memory. In: 2013 International Joint Conference on Neural Networks (IJCNN), pp. 1–8. IEEE (2013)
8. Hernandez, J., Abad, A.G.: Learning from multivariate discrete sequential data using a restricted Boltzmann machine model. In: 2018 IEEE 1st Colombian Conference on Applications in Computacional Intelligence (ColCACI), pp. 3450–3457. IEEE (2018)
9. Hinton, G.E.: Training products of experts by minimizing contrastive divergence. Neural Comput. **14**(8), 1771–1800 (2002)
10. Ji, S., Xu, W., Yang, M., Yu, K.: 3D convolutional neural networks for human action recognition. IEEE Trans. Pattern Anal. Mach. Intell. **35**(1), 221–231 (2013)
11. Laptev, I., Lindeberg, T.: Space-time interest points. In: Proceedings Ninth IEEE International Conference on Computer Vision, vol. 1, pp. 432–439, October 2003. https://doi.org/10.1109/ICCV.2003.1238378
12. Mnih, V., Larochelle, H., Hinton, G.E.: Conditional restricted Boltzmann machines for structured output prediction. arXiv:1202.3748 [cs, stat], February 2012
13. Nie, S., Ji, Q.: Capturing global and local dynamics for human action recognition. In: 2014 22nd International Conference on Pattern Recognition, pp. 1946–1951, August 2014. https://doi.org/10.1109/ICPR.2014.340
14. Salakhutdinov, R., Mnih, A., Hinton, G.: Restricted Boltzmann machines for collaborative filtering. In: Proceedings of the 24th International Conference on Machine Learning, ICML 2007, pp. 791–798. ACM
15. Schuldt, C., Laptev, I., Caputo, B.: Recognizing human actions: a local SVM approach. In: 2004 Proceedings of the 17th International Conference on Pattern Recognition, ICPR 2004, vol. 3, pp. 32–36, August 2004. https://doi.org/10.1109/ICPR.2004.1334462
16. Shu, N., Tang, Q., Liu, H.: A bio-inspired approach modeling spiking neural networks of visual cortex for human action recognition. In: 2014 International Joint Conference on Neural Networks (IJCNN), pp. 3450–3457. IEEE (2014)
17. Smolensky, P.: Information processing in dynamical systems: foundations of harmony theory (1986)
18. Sun, L., Jia, K., Yeung, D.Y., Shi, B.E.: Human action recognition using factorized spatio-temporal convolutional networks. In: Proceedings of the IEEE International Conference on Computer Vision, pp. 4597–4605 (2015)
19. Taylor, G.W., Hinton, G.E.: Factored conditional restricted Boltzmann machines for modeling motion style. In: Proceedings of the 26th Annual International Conference on Machine Learning, ICML 2009, pp. 1025–1032. ACM (2009)
20. Taylor, G.W., Hinton, G.E., Roweis, S.: Modeling human motion using binary latent variables. In: Advances in Neural Information Processing Systems, pp. 1345–1352. MIT Press (2007)
21. Veeriah, V., Zhuang, N., Qi, G.J.: Differential recurrent neural networks for action recognition. In: 2015 IEEE International Conference on Computer Vision (ICCV), pp. 4041–4049. IEEE (2015)

Application of Transfer Learning for Object Recognition Using Convolutional Neural Networks

Jesús Alfonso López Sotelo, Nicolás Díaz Salazar,
and Gustavo Andres Salazar Gomez(✉)

Automation and Electronics Department,
Universidad Autónoma De Occidente, Cali, Colombia
{jalopez, nicolas.diaz, gustavo.salazar}@uao.edu.co

Abstract. In this work, the transfer learning technique is used to create a computational tool that recognizes the objects of the automation laboratory of the Universidad Autónoma de Occidente in real time. As a pre-trained neural net, the Inception-V3 is used as a feature extractor in the images and on the other hand a softmax classifier is trained, this contains the classes that are going to be recognized. It was used Tensorflow platform with gpu in Python natively in Windows 10 and Opencv library for the use of video camera and other tools.

Keywords: Transfer learning · Softmax · Inception-V3 · Tensorflow
Neural networks · Convolutional neural network

1 Introduction

The conventional recognition of objects in images for a long time was carried out in a specific and imprecise way, because in order to detect the object and perform the recognition it was necessary to extract the characteristics manually, using different techniques depending on the task [1], techniques such as Principal Component Analysis (PCA), Independent Component Analysis (ICA), Local Binary Pattern (LBP), Wavelet transform, gabor transform, Histogram of oriented gradients (HoG), Scale-invariant feature transform (SIFT), among others. This caused that the system could not be used to recognize objects outside the context for which one of the aforementioned techniques had been tuned. With the arrival of deep learning, the feature extraction in images is now done automatically [2], thanks to convolutional neural networks (CNN), which allow the recognition of thousands of objects in images. in order to train these networks requires the use of a large number of images where the object that the network is going to recognize is located. A simple CNN can perform the task of recognizing objects in images with an accuracy of 57%, but in search of a better system, there are more complex networks with 80% accuracy. The training of more complex CNNs requires a high computing capacity, due to the number of images and neurons required by the network, and thanks to the Graphics Processor Unit (GPU) it is possible to do it more quickly [3], but it continues being a process that takes days and even weeks to complete, even if there are computers with exceptional capabilities. It is

A. D. Orjuela-Cañón et al. (Eds.): ColCACI 2018, CCIS 833, pp. 14–25, 2019.
https://doi.org/10.1007/978-3-030-03023-0_2

because of this that large companies and universities release to the public networks of complex and very precise architectures so that they can be used and evaluated by all those who need them. These networks allow users to use all their architecture leaving the last layer of classification free. re training this last layer is called transfer learning since the network is used as a feature extractor and the necessary classes of each user application are added, in this way the time and computation invested in the training is minimal, being few minutes, and providing that a computer with average characteristics is capable of executing these tasks. The objective of this project is to allow the use of pre-trained models for different applications without training a CNN from scratch, allowing many more people to make use of convolutional neural networks with mid-range computers. In Sect. 2, the generalities of convolutional neural networks are explained, this is because Sect. 3 talks about the neural network used for transfer learning, Inception-v3 and its architecture. In Sect. 4 we describe the concept of transfer learning to later talk about bottlenecks, what they are and what they are used for. Section 7 explains the classifier trained for this application and finally Sect. 9 shows the results of the tests in real time.

2 Convolutional Neural Networks

Convolutional neural networks are deep networks, which are composed of many interconnected neurons organized into many different layers [4], the specialty of these networks, as the name implies, is the operation "convolution", this is the equivalent of passing a filter by the image, where each neuron in the input layer passes a different filter, these convolution layers require that their neurons use a different activation function which is rectified linear unit (ReLU), which returns zero when values coming from the Neurons are negative and for the positive ones behaves in a linear way. The next layer is the max Pooling which chooses the highest values and reduces the dimensions of the layers as the network progresses. The last layer is the fully connected for which its activation function is the Softmax, it realizes the classification of all the data that comes out of the network and finally makes the prediction of the image, as shown in Fig. 1.

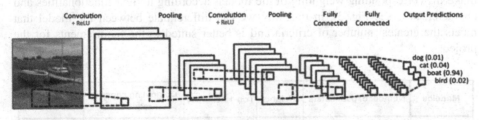

Fig. 1. Structure of a CNN [5].

3 Pre Trained Model

The selection of the pre-trained model is carried out for the development of this work, for which the most used models are weighted under the following criteria: Number of network parameters, network weight, top-1 precision and the computation capacity required for its training.

The first criterion is the number of parameters in the network, this gives an idea of the amount of calculations that the computer must perform to process an image through the entire network. The second criterion is the amount of space that the network occupies within the computer, where it is required that it is not elevated. The third criterion is the top-1 precision, which indicates how good the predictions of the model are, and it is intended to be as large as possible. The fourth and last criterion determines the amount of space in memory that requires the processing of an image inside the computer, seeking to use the least amount of this possible because there is a GPU of only 2 GB.

Based on the criteria described above, the following rating table is made (see Fig. 2), to select the most suitable pre-trained model for the development of this project. Scores of 1, 3 and 9 were assigned for the different comparison items.

Criteria Score	1	3	9
Number of parameters	Higher	Intermediate	Less
Weights	Higher	Intermediate	Less
Top-1 Accuracy	Higher	Intermediate	Less
Computing capacity required	Higher	Intermediate	Less

Fig. 2. Qualification of each criteria for pre-trained models.

According to the scores assigned to the respective selected criteria, we proceed to make the corresponding weighting of the models according to their functionalities and characteristics (see Fig. 3), in this way we can differentiate between the model that meets the greatest number of criteria and is better suited to the requirements for the project.

Modelos	Number of parameters	Weights	Top-1 Accuracy	Computing capacity required	Total
AlexNet	3	9	1	9	22
VGG-16	1	1	3	1	6
Inception-v3	9	3	9	3	24

Fig. 3. Weighting of alternatives of pre-trained models.

Figure 3 shows that the model with the highest score, which meets all the established requirements is Inception-v3, therefore, it is the model selected to work on this project.

4 Inception-V3

The Inception network is a network trained by Google, which has a top-1 accuracy of 78% and is capable of classifying 1000 categories. It is selected for this work due to its great acceptance in the community, easy use and high precision. The inception block is shown in the following (Fig. 4):

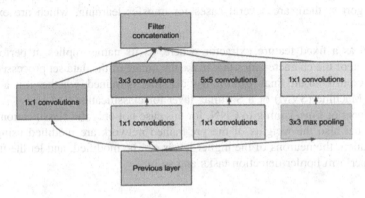

Fig. 4. Inception module

The block of the previous figure represents each module that has the Inception-V3, it is not the typical network that has layer after layer. The concept of the Inception is a microNetwork within another Network [6]; that is why at the exit of each layer there are 3 convolution layers (1 × 1, 3 × 3 and 5 × 5) and a max pooling operation, which will be concatenated and represent the output of that module, as well as the input of the next one. The complete architecture of the Inception-V3 network is shown in Fig. 5.

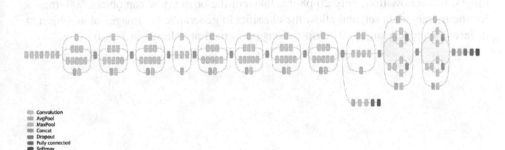

Fig. 5. Inception-V3 architecture [7].

As it can be observed, it has several processes in parallel of convolution followed by a concatenation stage, that is because it decreases the computational expense in a quarter [8] when the size of the grid of the convolutional stage is reduced, however, it is still almost impossible to train the 1000 categories, so the technique of transfer learning is used.

5 Transfer Learning

In practice, as already mentioned, training a CNN takes a lot of time and processing capacity, the latter is not easily accessible, so the transfer learning is a solution that has been received very quickly. The use of pre-trained models is a common practice, as some of these have been previously trained with more than 1,000,000 images with 1,000 categories, there are several cases for transfer learning, which are explained below:

- A CNN as a fixed feature extractor, where, as its name implies, it performs the extraction of the characteristics that make the image of the data set processed by the network and with the final result a new classifier is trained, whether it is a Support Vector Machine (SVM) or a Softmax layer for classification.
- Adjustment or Fine-Tuning to CNN, for this case not only the classification layer is trained but also the weights of the pre-trained network are modified using Back-propagation, the neurons of the higher levels can be modified, and let the first ones layers perform border detection tasks and so on.

6 Data Augmentation

The lack of images in unique categories, provide a possible over training in the network, guiding it to learn the categories or leaving the data set without enough images to perform a correct training. The data augmentation offers an alternative and a solution to these problems, with it you can generate an extensive data set with only process and apply different filters to a few images, increasing the number and diversity of these.

A script is developed that contains 24 operations and filters that are applied to a single image, these operations contain rotations, scaling, noise, filters, flips among others, this is how from only 20 photos taken of the object, you can obtain 500 images for the training data set, and allow the classifier to generalize the images of an object at different approach angles. The operations performed are shown in Fig. 6.

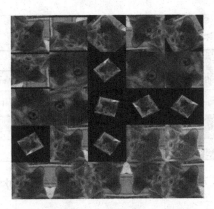

Fig. 6. Data augmentation operations.

7 Platforms

In this section, different deep learning platforms are compared in order to select the most appropriate for the implementation of the computational tool to be developed. It also describes the available hardware and the libraries used.

For the selection of a platform, different characteristics have been collected as shown in Fig. 7. The first criterion is that the platform is open source, thus reducing implementation costs for the tool and it has a large community that performs contributions to the software, which generates a great help for people who want to develop applications. As a second criterion, it is sought that the software can be installed in the largest number of operating systems to generalize the use of the tool. The third criterion is that the software has Python programming interface, since it is more popular around the world and provides a huge community for developments. On the other hand, Python is a high level language and very user friendly. For the fourth criterion it is required that the platform can perform the processing of the networks and their respective calculations in GPU to reduce training times. Finally, it is desired that the software allows the visualization of neural networks, in this way the architecture of the pre-trained models is known to perform the transfer learning, and it offers the possibility of carrying out a debugging process in case of errors in training. Scores of 1, 3 and 9 were assigned for the different comparison items.

According to the scores assigned to the respective selected criteria, we proceed to perform the corresponding weighting of the platforms according to their functionalities and characteristics, in this way we can differentiate between the platform that meets the most criteria appropriately.

Figure 8 shows that the platform with the highest score, which meets all the established requirements is TensorFlow, therefore, it is the software selected to work on this project.

Score			
Criteria	1	3	9
Open Source	No	-	Yes
Operative System	Less than 3	3	More than 3
Interface	Others	-	Python
GPU Use	No	Yes (Single-GPU)	Yes (Multi-GPU)
Network Visualization	No	-	Yes

Fig. 7. Rating of each criterion for frameworks.

Frameworks	Open Source	Operative System	Interface	GPU Use	Network Visualization	Total
Theano	9	3	9	3	9	33
Caffe	9	3	1	9	9	31
Torch	9	9	1	9	1	29
Tensor Flow	9	9	9	9	9	45
Microsoft CNTK	9	1	9	9	1	29
MatLab	1	3	1	3	1	9

Fig. 8. Weighting of alternatives for frameworks.

8 Bottlenecks

The bottlenecks are the term used in Tensorflow for the values of the output in the penultimate layer of a convolutional neural network [9], these have significant information of each class that we want to recognize, since the image is processed many times by the inception V3 due to its architecture. Each image is automatically adjusted to a standard size of 300 × 300 before being entered into the network, each time an image is processed, a * .txt file is created that has the same name of the image with its respective label. Once the entire dataset is processed by the network, a softmax classification layer is trained which has as input the * .txt files previously created, these contain the values generated from the last layer of the inception with a total of 2048 outputs, which are the characteristics that the network has extracted from the object and will be the inputs to the classification stage. The process is that of Fig. 9.

The dataset was created by the working group. The majority of images were taken from the automation laboratory of the Universidad Autónoma de Occidente, with a total of 27 classes and 500 images per class, the dataset contains 13500 images in total.

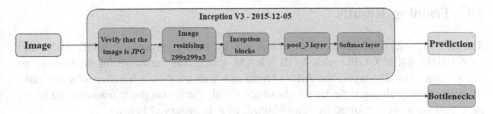

Fig. 9. Bottleneck's creation process.

9 Classification

The sorting process starts by loading all the bottlenecks into a vector and coding the outputs depending on the name of the bottleneck. For this work, we have a total of 27 classes, which are indicated below (Fig. 10):

1. Gato	14. Boligrafo
2. Perro	15. PLC
3. Carro	16. Mesa_xy
4. Basura	17. Toma Corriente
5. Ágora	18. Labvolt
6. Laptop	19. Cilindro
7. Celular	20. Silla
8. Teclado	21. Planta de Presión
9. Mouse	22. Planta Piloto
10.CPU	23. Servo motor
11.Monitor	24. Osciloscopio
12.Gafas	25. Multímetro
13.Cuaderno	26. Fuente
	27. Generador de señal

Fig. 10. Network classes.

This allows the code to load all the bottlenecks with their respective names and be able to perform the training. Once all the features configured in the code have been met, the address of the bottlenecks is verified, and the code for the training is carried out, which performs 10 epochs. Each of these indicates that the entire dataset has been processed, therefore, upon completion of the training of the classifying layer, it has processed 10 times the entire data set of the images.

Finally, this classify layer is stored inside the folder where the code is located, with the name of graph because it is called networks in the context of Tensorflow, and it is saved with extension * .pbtxt (protocol buffer, format that works the platform). The protocol buffer are data structures defined in text files, which generate classes in different programming languages. These classes can be easily loaded, saved and accessed. The saver class of Tensorflow is used to create graph of the network. The complete process for training the classifier corresponds to entering the dataset in Inception-v3, obtaining the bottlenecks. With these files the class coding is done and the classifier is finally trained.

10   Training Results

The training was done on an Asus laptop with an Intel I5 processor, 8 GB RAM, and an NVIDIA 940MX GPU with 2 GB VRAM. Using this hardware for the training of the softmax classify layer, the Adam Optimizer was used as a function of error minimization, the training time lasted about 27 s with these computer features, the accuracy and error were plotted in Tensorboard, as it is observed below:

As shown in Fig. 11, the accuracy of the layer increased, reaching a value of 0.9985, which indicates that the training was successful.

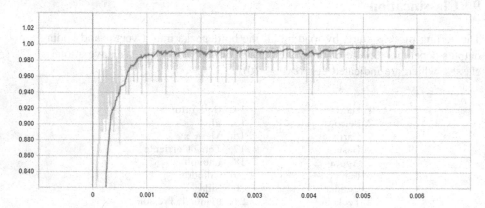

Fig. 11. Training network accuracy.

A confusion matrix was also made with a batch of data for evaluation, which were not present in the training, the numbers in the labels of the confusion matrix are associated with the class numbers in the classification section. In order to elaborate this matrix a batch of 1619 data was taken, so that to make the predictions with the network already trained, and 1619 labels to be able to compare the results. The network had 26 misclassified objects indicating a precision of 98.39%.

11   Real Time Test

When the graph has been saved successfully, it can be used to make predictions in the selected environment, in the first instance the red is loaded in the new code, indicating the name and address of the file, a session is created in Tensorflow with the sorting network and also loading the inception-v3, the connection with the camera and its respective port is made and the names of the classes that are going to be displayed on the screen are added.

Once these constant characteristics have been defined within the code, a loop is entered to process and predict the images taken by the camera until the 'q' key is pressed and the code is stopped. Each frame taken by the camera is processed and the network makes the prediction giving a score to each class, these scores are ordered from highest to lowest, and a threshold is added which is at 85% accuracy, so if the higher score does not exceed this value will not be taken into account and the highest or none is printed on the screen, this indicates the object that is in the image, the complete process to make a prediction on images in real time is evidenced in the Fig. 12. The processing time by the network and the new classifier is 0.166 s per image, which translates to 6 frames per second (FPS).

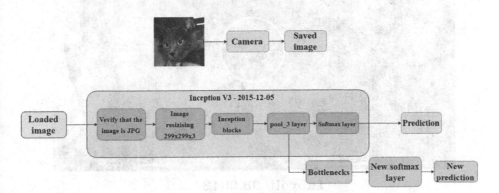

Fig. 12. Prediction process for an image.

The following images are screenshots of the network in operation, making the predictions of each picture (Figs. 13 and 14).

Planta_presion 99.9988

Fig. 13. Real time test with the "Planta de presión".

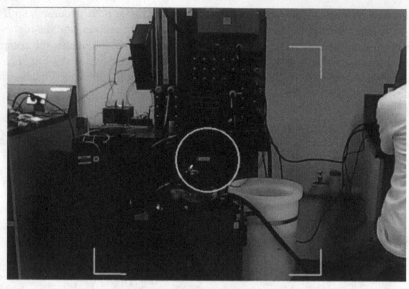

Labvolt 98.0042

Fig. 14. Real time test with the "planta LabVolt".

12   Conclusions

In this project the transfer learning technique was developed in order to use a convolutional neural network trained as a base for the recognition of new chosen objects within a work area. CNNs greatly facilitate the extraction of features within the image, having the ability to make known the aspects that make an object is to another, and thanks to these data, to make a simple classifier can differentiate between different categories. The applications of this technique range from recognition and differentiation of certain objects to the diagnosis of medical images with the proper processing of them. Currently, CNNs have acquired great importance in the processing of images, since they allow their automatic analysis, leading to great technological advances in a large part of the areas of science. One of the achievements of this project with the use of these networks was to be able to reach more people and use them in many more areas, facilitating their understanding and training.

The transfer learning is a technique that has facilitated the access of many people to the most precise convolutional networks that currently exist, this generates that any type of applications can be made without the need of having a high-end computer and therefore a high cost in hardware for training. In addition, it greatly reduces the time needed to train a network of these, because the pre-trained model used in this project, which acts as a feature extractor, takes approximately 30 min to process 13500 images and the MLP that makes as classifier takes approximately 20 s to complete all training with the bottlenecks. One of the objectives achieved with this project is to demonstrate the ease of handling a pre-trained network, and decrease hardware and software costs for a specific application in the use of CNN, which is achieved as the only training carried out is the MLP classification. For better and more optimal results it is recommended to use the state of the art in algorithms such as the Adam Optimizer, and to use the correct functions for the type of network, as is the example of the crossed entropy error that is the best in terms of tasks of classification for MLP networks.

The use of CNN has increased significantly over the last few years, allowing each time the new generations of networks to be more precise, even surpassing human capacity in image recognition.

Some applications that can be achieved with the technique described in this paper is the use of convolutional networks to help geologists in field work, sometimes it is difficult to discern what type of rock is found and it is not always possible to do laboratory tests to be sure. However, the geologist could use an application that supports their criteria and is able to define the type of rock that is in front. This with a dataset that has most types of rocks. On the other hand, it is possible to re-train a convolutional network with thermographic images and use it to carry out predictive maintenance in industrial plants or different facilities. In addition, it can be used for traffic analysis with classes that are not found in other countries like "motocarro", in this way you can expand the range of vehicles that can be recognized by the traffic cameras in Colombia.

References

1. Castrillon, W., Alvarez, D., López, A.: Técnicas de extracción de características en imágenes para el reconocimiento de expresiones faciales. In: Scientia Et Technica, vol. XIV, no. 38, pp. 7–12 (2008)
2. Zhou, B., Khosla, A., Lapedriza, A., Oliva, A., Torralba, A.: Object detectors emerge in deep scene CNNs (2015)
3. Brown, L.: Deep learning with GPUs (2015)
4. Loncomilla, P.: Deep learning: redes convolucionales (2016)
5. The data science blog. An intuitive explanation of convolutional neural networks. https://goo.gl/KdqfLV
6. Szegedy, C., et al.: Going deeper with convolutions (2014)
7. Alemi, A.: Improving inception and image classification in tensorflow (2016)
8. Szegedy, C., Vanhoucke, V., Ioffe, S., Shlens, J., Wojna, Z.: Rethinking the inception architecture for computer vision (2015)
9. Tensorflow how to retrain inception's final layer for new categories. https://www.tensorflow.org/tutorials/image_retraining#bottlenecks

SOM-Like Neural Network and Differential Evolution for Multi-level Image Segmentation and Classification in Slit-Lamp Images

Hans Israel Morales-Lopez[✉], Israel Cruz-Vega, Juan Manuel Ramirez-Cortes,
Hayde Peregrina-Barreto, and Jose Rangel-Magdaleno

Instituto Nacional de Astrofísica, Óptica y Electrónica, Apartado Postal 51-216,
72840 Puebla, Mexico
hans.israel@inaoep.mx

Abstract. A nuclear cataract is a type of disease of the eye that affects a considerable part of the human population at an advanced age. Due to the high demand for clinical services, computer algorithms based on artificial intelligence have emerged, providing acceptable aided diagnostics to the medical field. However, several challenges are yet to be overcome. For instance, a well-segmented image of the region of interest could prove valuable at a previous stage in the automatic classification of this disease. A great variety of research in image classification uses several image processing techniques before the classification stage. In this paper, we explore the automatic segmentation based on two leading techniques, namely, a Self-Organizing Multilayer (SOM) Neural Network (NN) and Differential Evolution (DE) methods. Specifically, the fuzzy entropy measure used here is optimized via a neural process, and by using the evolutive technique, optimal thresholds of the images are obtained. The experimental part shows significant results in getting a useful automatic segmentation of the medical images. In this extended version, we have implemented the use of a Multilayer Perceptron, a classifier that proves the usefulness of the segmented images.

Keywords: SOM · Neural network · Image segmentation
Differential evolution · Multilayer perceptron · Classification

1 Introduction

The segmentation of images by defining regions of interest has an essential role in imaging applications [6]. The segmentation process subdivides an image into its constituent parts or objects according to specific properties. Carrying the level of subdivision depends on the problem under analysis and the characteristics of the image itself. In other words, segmentation is achieved when the objects of interest in an application have been isolated [13]. Segmentation is often the first

© Springer Nature Switzerland AG 2018
A. D. Orjuela-Cañón et al. (Eds.): ColCACI 2018, CCIS 833, pp. 26–37, 2018.
https://doi.org/10.1007/978-3-030-03023-0_3

stage in pattern recognition systems. Once the objects of interest are separated from the rest of the image, specific characterizing measurements are usually realized (*feature extraction*). After that, these are used to *classify* the objects into particular groups or classes [5].

The majority of the segmentation algorithms based on thresholding produce a two-level, or "object-background" segmentation. Oftentimes, this is unsatisfactory for applications where several objects or classes need to be detected [3]. A general classification of segmentation techniques is either contextual or non-contextual. *Non-contextual techniques* ignore the relationships between features in an image; some global attribute simply groups the pixels. On the other hand, *contextual techniques* exploit the relationships between image features [5].

In most of the medical applications, image segmentation is vital due to the interest of subdividing images into several regions that can *label* tissue affected by some disease, as opposed to healthy tissue.

In this paper, we propose a comparison of multi-level segmentation using differential evolution and a SOM-like neural network (SLNN). For the case of the SLNN, we use labels that are automatically pre-selected by a fuzzy clustering technique. The images of the eye obtained by both techniques come from a slit-lamp. The rest of the paper is divided as follows: Sect. 2 describes some related work concerning the segmentation problem. In Sect. 3 the theory related to the multi-level segmentation techniques is explained, and a brief description of the Multilayer Perceptron is presented. Section 4 shows the experimental results. And finally, the conclusion of this work is presented in Sect. 5.

2   Related Work

There are many related works to image segmentation techniques. Kapur *et al.* (1985), Wong and Sahoo (1989), Pal (1996) and Rosin (2001) used entropy-based global-thresholding for image segmentation. Sarkar *et al.* (2015) proposed a multi-level thresholding approach based on fuzzy partition of the image histogram and entropy theory [14].

A variety of methods have been developed to solve the problem of image segmentation, which is an essential stage in an automatic diagnosis system. Cellular Neural Networks (CNN) proved to be very useful regarding real-time image processing. For instance, in [17], the Discrete-Time Cellular Neural Networks (DTCNN) are applied to image segmentation based on active contour techniques. Due to the iterative action of the algorithm, parallel processing is only partially ensured in specific steps. Aizenberg *et al.* [1] present a particular kind of cellular neural network based on a multiple-valued threshold logic in the complex plane. The use of a CNN based on Universal Binary Neurons (UBNs) makes the computation very efficient since it only requires boolean operations. Vilariño *et al.* present a new algorithm for the cellular active contour technique in [16]. This algorithm has a high efficiency and flexibility concerning the limits of interest pursued.

A three-level thresholding method for image segmentation is presented in [15]. It is based on probability partition, fuzzy partition, and on the entropy theory.

The fuzzy entropy, as a measure of error, is defined through probability analysis. Also, the image is divided into three parts, namely, dark, gray and white. The procedure for finding the optimal combination of all the fuzzy parameters is implemented by a genetic algorithm with an appropriate coding method to avoid useless chromosomes. Awad *et al.* [2] presents a new multicomponent image segmentation method based on a nonparametric unsupervised artificial neural network called Kohonen's self-organizing map (SOM) and on a hybrid genetic algorithm (HGA). SOM is used to detect the main features in the image. After that, HGA optimizes the clusters of the image into homogeneous regions without any prior knowledge. An unsupervised parallel segmentation approach that uses a fuzzy Hopfield neural network (FHNN) is presented in [10]. Its primary purpose is to embed fuzzy clustering into neural networks, so that online learning and parallel implementation for medical image segmentation are feasible. A generation of possible results follows a fuzzy c-means clustering strategy that is included in the Hopfield neural network to eliminate the need to find weighting factors in the energy function. The formulation is based on a fundamental concept commonly used in pattern classification, called the "within-class scatter matrix" principle.

It is notorious in these references that most of the segmentation techniques are based on unsupervised learning with Shannon Entropy as a measure of error. The main difference between fuzzy entropy and Shannon entropy is that fuzzy entropy contains fuzzy (possibilistic) uncertainties, whereas Shannon entropy contains random uncertainties (probabilistic).

3   Self-organizing Maps, Differential Evolution and Multilayer Perceptron

This section explains the two main techniques addressed for the segmentation procedure, namely, the SOM-like neural network (SLNN) and Differential Evolution (DE). However, we first lay the foundations to represent the image related to fuzzy concepts, enabling it to be processed by the neural network and the evolutionary algorithm.

3.1   Histogram Based Image Fuzzification

An image I of size $M \times N$ with L gray intensity levels may be seen as an array of fuzzy singletons whose value of membership denotes the degree to which they possess some property or feature [3]. Therefore, a gray level image can be interpreted as a fuzzy C partition, as shown in Fig. 1, using the fuzzy C-means algorithm [8].

3.2   Error Function Definition

In terms of the fuzzy set theory, the fuzziness measures of fuzzy partitions are usually used as a validation measure of a clustering solution. They measure the

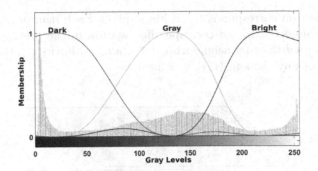

Fig. 1. Histogram based fuzzification

fuzziness of a partition of data set [3]. According to [4], the fuzziness measure of a fuzzy partition should satisfy the following properties:

- $F(A) = 0$ (or other unique minimum) if $\mu_A(x_i) = 0$ or 1, $\forall i$ (if A is crisp);
- $F(A) =$ a unique maximum if $\mu_A(x_i) = 0.5$, $\forall i$;
- $F(A) \geq F(A^*)$ where A^* is a sharper version of A, i.e.,

$$\left(\begin{array}{l} \mu_A^*(x_i) \leq \mu_A(x_i) \text{ for } \mu_A(x_i) \leq 0.5 \\ \mu_A^*(x_i) \geq \mu_A(x_i) \text{ for } \mu_A(x_i) \geq 0.5 \end{array} \right);$$

- $F(A) = F(A^C)$, where A^C is the complement of A.

where

$F(A)$ fuzziness measure of fuzzy set A;
μ membership function in the fuzzy set;
x_i supporting points of the fuzzy set A.

An error function can be defined, satisfying the previously described properties, as logarithmic entropy [4]:

$$H(A) = -\frac{1}{n \ln(2)} \sum_i \Big[\mu_A(x_i) \ln \mu_A(x_i) + (1 - \mu_A(x_i))(1 - \ln \mu_A(x_i)) \Big] \quad (1)$$

where n is the number of fuzzy sets.

3.3 Self-organizing Multilayer Neural Network [9]

As it is, the multilayer perceptron (MLP) cannot be used for image segmentation when only one input image is available because the MLP requires more than one image for the learning process [8]. For the one-image segmentation problem, it is desirable to apply a self-supervised learning technique, notably one which is auto-adaptive (or self-organizing) [3].

The network consists of an input layer, an output layer, and one hidden layer because of the training process (Fig. 2). Each layer consists of $M \times N$ neurons, where every neuron corresponds to an image pixel. Each neuron in the current layer is only connected to the corresponding neuron in the previous layer and the neurons in its d-th order neighborhood. For a $m \times n$ lattice (l), the d-th order neighbor, N_{ij}^d of any element (i, j) is defined as

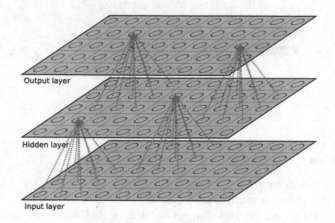

Output layer

Hidden layer

Input layer

Fig. 2. SOM-like NN architecture

$$N_{ij}^d = \{(i, j) \in l\} \tag{2}$$

such that

- $(i, j) \notin N_{ij}^d$
- if $(m, n) \in N_{ij}^d$, then $(i, j) \in N_{mn}^d$

(m, n) denotes the location of the pixels surrounding the (i, j) central pixel. As an example, $N^1 = \{N_{ij}^1\}$ can be obtained by taking the four nearest-neighbor pixels. $N^2 = \{N_{ij}^2\}$ consists of the eight pixels neighboring (i, j) [8]. As a condition, the NN's weights cannot be randomly set, or they will alter the input image [3]. Random initialization of weights may act as pseudo noise, and the compactness of the extracted regions may be lost [7].

The activation function of every neuron consists of a sigmoid-like function with multiple levels, described in Eq. (3). The activation function is a key part in this NN because the output needs to be set in the levels given by this function. It is obtained by the superposition of $k = C - 1$ shifted sigmoid functions (C being the number of classes obtained by the fuzzification).

$$f(x) = \sum_k \left(\frac{y_k - y_{k-1}}{1 + e^{-\frac{x - \theta_k d}{\theta_0}}} \right) \times \left[u \left(x - y_{k-1} \times d^2 \right) - u \left(x - y_k \times d^2 \right) \right] \tag{3}$$

where

θ_k thresholds;
y_k target level of each sigmoid;
θ_0 steepness parameter;
 d size of the neighborhood;
 u step function.

The thresholds and the target values are obtained from the error function, as the gray levels with the maximal and minimal levels of fuzziness, respectively [3].

3.4   Multi-level Fuzzy Entropy

In terms of the fuzzy sets theory, when a data set is given, uncertainty is related either to vague descriptions or inaccurate measurements, and it is called the fuzziness measure [12]. The fuzziness of a fuzzy set is a measure of the information contained in it, just like probabilistic entropy is in the field of information theory [3]. Entropy was chosen as a measure of error because of the learning process: the aim of the network is to reduce the degree of fuzziness of the input image. Fuzzy entropy, when used for multi-level segmentation, measures the amount of information extracted [18]. According to the fuzzy theory, a fuzzy set is a generalization of a classical set, where an element may partially belong to a set A. Therefore, A can be defined as:

$$A = \{(x, \mu_A(x)) \,|\, x \in X\} \tag{4}$$

where $\mu_A(x)$ is called the membership function, which defines the closeness of x to A. The fuzzy entropy for each segment can be defined by:

$$H_n(t) = -\sum_{i=0}^{L-1} \frac{p_i \times \mu_n(i)}{P_n} \ln \frac{p_i \times \mu_n(i)}{P_n} \tag{5}$$

where

$$P_n = \sum_{i=0}^{L-1} p_i \times \mu_n(i) \tag{6}$$

3.5   Differential Evolution (DE)

It is a population-based global optimization algorithm [14]. The i-th individual of the population at a generation t is a D-dimensional vector containing a set of D optimization parameters:

$$\boldsymbol{Z}_i(t) = [Z_{i,1}(t), Z_{i,2}(t), \dots, Z_{i,D}(t)] \tag{7}$$

In each generation, a *donor* vector $Y_i(t)$ is created. To create a donor vector for each i-th member, three other parameter vectors $Z_{r1,j}$, $Z_{r2,j}$ and $Z_{r3,j}$ are randomly chosen from the current population. The donor vector is obtained by multiplying a scalar number F by the difference between any two of the three parameter vectors. The process for the j-th component of the i-th vector is:

$$Y_{i,j}(t) = Z_{r1,j}(t) + F\left(Z_{r2,j}(t) - Z_{r3,j}(t)\right) \tag{8}$$

A crossover operation is performed to increase the diversity of population. It takes place on each of the D variables whenever a randomly picked number between 0 and 1 is within the Cr value. Thus, for each target vector $Z_i(t)$, a trial vector $R_i(t)$ is created as follows:

$$R_{i,j}(t) = \begin{cases} Y_{i,j}(t) \text{ if } \mathrm{rand}_j(0,1) \leq Cr \text{ or } j = rn(i) \\ Z_{i,j}(t) \text{ otherwise} \end{cases} \tag{9}$$

Where $j = 1, 2, \ldots, D$, $\mathrm{rand}_j(0,1) \in [0,1]$ is the j-th evaluation of a generator of uniform random numbers and $rn(i) \in [1, 2, \ldots, D]$ is a random index that ensures that $R_I(t)$ contains at least a component from $Z_I(t)$. Cr is the crossover probability [12].

Finally, selection is performed to determine which one vector will survive among the target and trial vectors in the next generation $t = t + 1$:

$$Z_I = \begin{cases} R_I(t) & f\left(R_I(t)\right) > f\left(Z_I(t)\right) \\ Z_I(t) & f\left(R_I(t)\right) \leq f\left(Z_I(t)\right) \end{cases} \tag{10}$$

where f is the function to be maximized.

Finally, Fig. 3 presents the flowcharts for SOM-like multi-level segmentation and DE multi-level segmentation, respectively.

3.6 Multilayer Perceptron

A multilayer perceptron (MLP) is a class of feed-forward artificial neural network. An MLP consists of at least three layers of nodes. Except for the input nodes, each node is a neuron that uses a nonlinear activation function. MLP utilizes a supervised learning technique called error back-propagation for training. It can distinguish data that is not linearly separable. Since MLPs are fully connected, each node in one layer connects with a certain weight w_{ij} to every node in the following layer [11]. The two common activation functions are both sigmoids, and they are described by:

$$y(x_i) = \tanh(x_i) \tag{11}$$

and

$$y(x_i) = \frac{1}{1 + e^{-x_i}} \tag{12}$$

The first is a hyperbolic tangent that ranges from -1 to 1, whereas the other one is the logistic function, which is similar in shape, but ranges from 0 to 1. Here y_i is the output of the i-th node (neuron), and x_i is the weighted sum of the input connections.

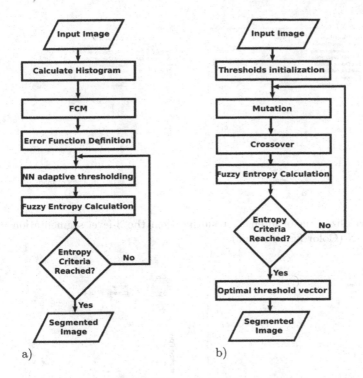

Fig. 3. Flowcharts for (a) SOM-like and (b) DE multi-level segmentation

Learning occurs in the perceptron by changing the connection weights after each piece of data is processed, based on the amount of error in the output compared to the expected result. This is an example of supervised learning, and is carried out through error's back-propagation, a generalization of the least mean squares algorithm in the linear perceptron.

4   Experimental Results

The dataset consists of 40 images, labeled either as cataract or non-cataract. The images are obtained from a digital slit-lamp camera. The images shown are 1028×640 RGB images; they were focused with a Topcon SL-D2 slit-lamp, using a slit angle of $30°$, a 0.5 mm aperture, and a maximum luminous intensity. They were captured by a Topcon DC-1 digital camera. For both multi-level segmentation techniques, the maximum number of iterations was set to 1000, achieving good stabilization for both methods (DE and SLNN). The number of thresholds was set to 2, 3 and 4 in order to test the multi-level segmentation and classification. The effectiveness of the segmentation techniques was measured in fuzzy entropy terms.

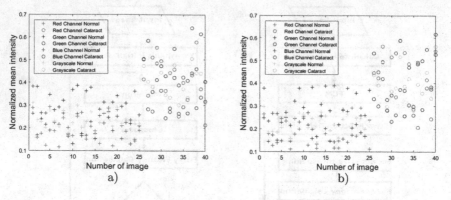

Fig. 4. Normalized mean intensity resulting from the 3-level segmentation using (a) DE (b) SLNN (Color figure online)

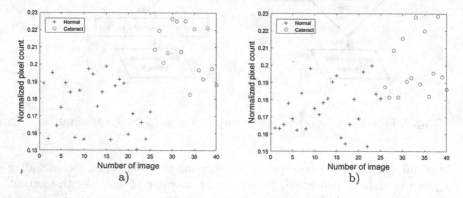

Fig. 5. Normalized pixel count resulting from the 3-level segmentation using (a) DE (b) SLNN (Color figure online)

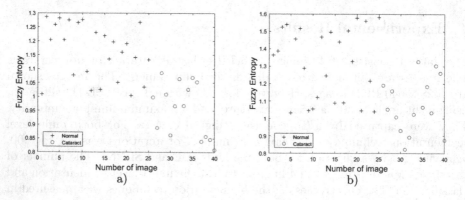

Fig. 6. Fuzzy entropy of the 3-level segmented image using (a) DE (b) SLNN (Color figure online)

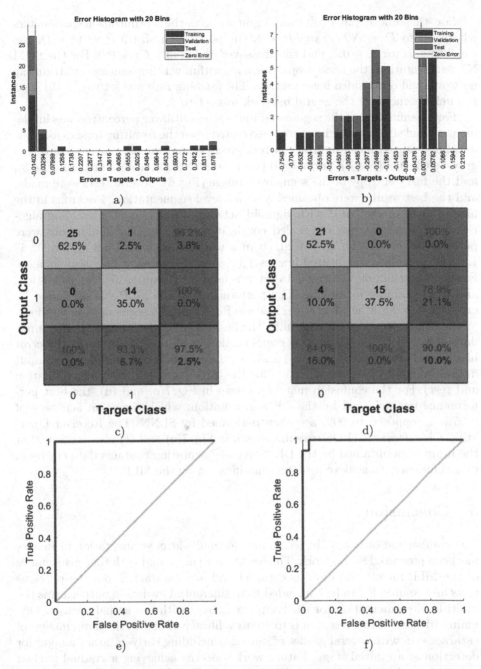

Fig. 7. Results of the training; Error using (a) DE, (b) SLNN; Confusion matrix for training, validation and testing using: (c) DE, (d) SLNN; ROC for training, validation and testing using: (e) DE, (f) SLNN

For the differential evolution segmentation, the optimization parameters where set to $D = NThresholds \times 2$, the population size $NP = 10 \times D$, the weighting factor $F = 0.5$, and the crossover probability $Cr = 0.9$. For the SOM NN segmentation, the back-propagation algorithm was implemented to train the network, and one hidden layer was set. The learning rate was set to $\eta = 0.1$, and the neighborhood of the neural network was set to 3.

For classification of the segmented images, a multilayer perceptron was implemented, and six characteristics were extracted from the resulting images as input data: normalized mean intensity in red, green, blue channels and in a converted grayscale image (Fig. 4), normalized pixel count of the segmented region (Fig. 5), and the fuzzy entropy of the segmented image (Fig. 6). Several tests were made, and the best results were obtained with a 3-level segmentation, 4 neurons in the hidden layer of the MLP with sigmoid activation function. The training algorithm implemented was the scaled conjugated gradient. The simulations were performed with MATLAB® R2017b in a workstation with Intel® Core™ i7 1.8 GHz processor. Segmented images $f_s(x, y)$ were formed as gray scale images. For classification, 60% of the data set was used for training, 15% for validation, and the remaining for testing. The performance of the MLP was measured by cross entropy, and the results are shown in Fig. 7 for both segmentation methods.

As can be observed in the results, the best training was achieved by the input data obtained by the DE 3-level segmentation in all possible ways. In the error histogram bars shown in Fig. 7(a) and (b), the error was minimized by the input data obtained by DE segmentation for the three data sets (training, validation and test). For the confusion matrices shown in Fig. 7(c) and (d), the best performance was achieved by the DE segmentation, which shows an accuracy of 97.5%, as opposed to 90% accuracy performed by SLNN. The Receiver Operating Characteristic (ROC) results shown in Fig. 7(e) and (f) also confirm that the input data obtained by the DE 3-level segmentation contains data clustered more efficiently to achieve a better classification by the MLP.

5   Conclusion

The comparison of two main techniques for multi-level segmentation of images has been presented in this work. The results were measured with the performance of the MLP for classification as cataract and non-cataract. From the previous tests and results, it can be concluded that differential evolution outperforms the multi-level segmentation for slit-lamp eye images with the actual dataset. One main objective of this research is to create a library of digital slit-lamp images of cataract eyes with several grades of damage, including early-cataract images for detection at an initial stage. Future work concerns achieving a gradual nuclear cataract classification of slit-lamp images and using deep networks for the same purpose.

References

1. Aizenberg, I., Aizenberg, N., Hiltner, J., Moraga, C., Zu Bexten, E.M.: Cellular neural networks and computational intelligence in medical image processing. Image Vis. Comput. **19**(4), 177–183 (2001)
2. Awad, M., Chehdi, K., Nasri, A.: Multicomponent image segmentation using a genetic algorithm and artificial neural network. IEEE Geosci. Remote. Sens. Lett. **4**(4), 571–575 (2007)
3. Boskovitz, V., Guterman, H.: An adaptive neuro-fuzzy system for automatic image segmentation and edge detection. IEEE Trans. Fuzzy Syst. **10**(2), 247–262 (2002)
4. De Luca, A., Termini, S.: A definition of a nonprobabilistic entropy in the setting of fuzzy sets theory. Inf. Control. **20**(4), 301–312 (1972)
5. Dougherty, G.: Digital Image Processing for Medical Applications. Cambridge University Press, Cambridge (2009)
6. Gacsádi, A., Szolgay, P.: Variational computing based segmentation methods for medical imaging by using CNN. In: 2010 12th International Workshop on Cellular Nanoscale Networks and Their Applications (CNNA), pp. 1–6. IEEE (2010)
7. Ghosh, A.: Use of fuzziness measures in layered networks for object extraction: a generalization. Fuzzy Sets Syst. **72**(3), 331–348 (1995)
8. Ghosh, A., Pal, N.R., Pal, S.K.: Self-organization for object extraction using a multilayer neural network and fuzziness mearsures. IEEE Trans. Fuzzy Syst. **1**(1), 54–68 (1993)
9. Gurney, K.: An Introduction to Neural Networks. CRC Press, London (2014)
10. Lin, J.-S., Cheng, K.-S., Mao, C.-W.: A fuzzy hopfield neural network for medical image segmentation. IEEE Trans. Nucl. Sci. **43**(4), 2389–2398 (1996)
11. Pal, S.K., Mitra, S.: Multilayer perceptron, fuzzy sets, and classification. IEEE Trans. Neural Netw. **3**(5), 683–697 (1992)
12. Paul, S., Bandyopadhyay, B.: A novel approach for image compression based on multi-level image thresholding using shannon entropy and differential evolution. In: 2014 IEEE Students' Technology Symposium (TechSym), pp. 56–61. IEEE (2014)
13. Gonzalez, R.C., Woods, R.: Digital Image Processing. Pearson Education, London (2002)
14. Sarkar, S., Paul, S., Burman, R., Das, S., Chaudhuri, S.S.: A fuzzy entropy based multi-level image thresholding using differential evolution. In: Panigrahi, B.K., Suganthan, P.N., Das, S. (eds.) SEMCCO 2014. LNCS, vol. 8947, pp. 386–395. Springer, Cham (2015). https://doi.org/10.1007/978-3-319-20294-5_34
15. Tao, W.-B., Tian, J.-W., Liu, J.: Image segmentation by three-level thresholding based on maximum fuzzy entropy and genetic algorithm. Pattern Recognit. Lett. **24**(16), 3069–3078 (2003)
16. Vilariño, D.L., Rekeczky, C.: Pixel-level snakes on the CNNUM: algorithm design, on-chip implementation and applications. Int. J. Circ. Theory Appl. **33**(1), 17–51 (2005)
17. Vilariño, D.L., Cabello, D., Pardo, X.M., Brea, V.M.: Cellular neural networks and active contours: a tool for image segmentation. Image Vis. Comput. **21**(2), 189–204 (2003)
18. Yin, S., Zhao, X., Wang, W., Gong, M.: Efficient multilevel image segmentation through fuzzy entropy maximization and graph cut optimization. Pattern Recognit. **47**(9), 2894–2907 (2014)

Implementation of a Neural Control System Based on PI Control for a Non-linear Process

Diego F. Sendoya-Losada[✉] ⓘ, Diana C. Vargas-Duque ⓘ,
and Ingrid J. Ávila-Plazas ⓘ

Department of Electronic Engineering, Faculty of Engineering,
Surcolombiana University, Neiva, Huila, Colombia
diego.sendoya@usco.edu.co

Abstract. This paper explores the possibility of using a machine learning algorithm such as artificial neural networks to control a non-linear liquid level system. To achieve this objective, PI controllers were designed for two different scenarios: In the first, a single PI controller was used to control the system at one setpoint. In the second, 4 PI controllers were designed in order to cover a wider operating range of the plant. The input and output signals from the PI controllers were used to train a controller based on artificial neural networks. The neural network that presented greater simplicity and lower computational cost was selected. In this case, a neural network with 3 hidden layers and 20 neurons per layer was the one that best recreated the dynamics of the PI controllers. The root-mean-square error (RMSE) was used to validate the results obtained with the PI controllers and with the controller based on neural networks. In both scenarios the variations of the error were smaller when the neuronal controller was used than when the PI controllers were used. The results show that machine learning algorithms such as artificial neural networks can be used effectively to control processes whose dynamics are complex.

Keywords: Artificial neural network · Machine learning · Neural control
PI controller

1 Introduction

In almost all the applications of industrial processes, the control of the variables is critical for the safe and efficient operation of the same. The most common controlled variables are pressure, level, temperature and flow. Level control loops are very common in the industry, in fact, they occupy the second place after the flow control loops. Due to the importance and the large number of processes that require a precise level control, the Surcolombiana University recently acquired a tank system called CE105 MV. This system presents a configuration similar to that which can be found in many industrial applications or as part of a much larger and more sophisticated plant.

Usually the control of the liquid level in a tank system is done by classical control techniques such as PI or PID, due to its well-known and simple structure. For the design of these conventional controllers it is necessary to choose a setpoint and then to find a linear model of the system, ensuring that the control works well in this region,

© Springer Nature Switzerland AG 2018
A. D. Orjuela-Cañón et al. (Eds.): ColCACI 2018, CCIS 833, pp. 38–49, 2019.
https://doi.org/10.1007/978-3-030-03023-0_4

but when it moves away from the setpoint, the controller loses effectiveness. When conventional controls for non-linear processes are designed, it is necessary to develop controllers for each setpoint of the system. Furthermore, if the dynamics of the plant are too complex the design of many PI controllers will be required. Therefore, one of the main advantages with the design of a single controller based on neural networks is that it allows to replace all these PI controllers in a single controller that guarantees the good functioning of the system regardless of the non-linearities of the process.

The machine learning algorithms can lead to significant advances in the automatic control of this type of systems. A machine learning algorithm that has the potential to accomplish this is the artificial neural network (ANN). Neural networks are actually a pretty old idea, but they fell into disuse for a while. There are many papers addressed the PID, fuzzy or neural networks control in the water or liquid level control system. In [1], an ANN is utilized as advanced process control technique for water treatment. An adaptive model reference fuzzy controller is presented in [2], for controlling the water level in a water tank. A back propagation neural network algorithm is used to adjust the parameters of the PID controller and control the liquid level of molten steel smelling non-crystalloid in [3]. Nowadays, although the ANN is a cutting-edge technique for many different machine learning problems, it is important to expand its use to other areas such as the industrial process control. In recent years, advances in machine learning algorithms and hardware have led to more efficient methods for the training of artificial neural networks [4]. Neural networks are parameterized as nonlinear functions. Their parameters are, for instance, the weights and biases of the network. Adjustment of these parameters results in different shaped nonlinearities. Typically, the adjustment of the neural network parameters is achieved by a gradient descent approach on an error function that measures the difference between the output of the neural network and the output of the actual system (function). That is, it is tried to adjust the neural network to serve as an approximation for an unknown function that it is only known by how it specifies output values for the given input values (i.e., the training data). This document focuses on presenting the possibility of using an machine learning algorithm such as ANN, to control the level of liquid in a tank whose dynamics are non-linear. With this study, it was pretended to mimic the PI controller using an ANN algorithm. The simulation is done with Matlab/Simulink. The detailed results of the comparison study between the neural controller and the PI controller are presented to demonstrate the performance and effectiveness of the proposed algorithm.

2 Materials and Methods

2.1 Mathematical Model of the System

The mathematical model presented is given by the equations that describe the complete system, in this case it is a non-linear system. Figure 1 shows its respective scheme. The dynamic model is determined by the relationship between the input flow $q_i(t)$ and the output flow $q_o(t)$ through the valve. Equation (1) describes this relationship.

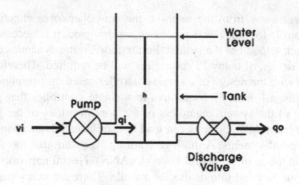

Fig. 1. System diagram.

$$q_i(t) - q_o(t) = \dot{V}(t) = A\dot{h}(t) \tag{1}$$

$V(t)$ is the volume of the tank, A is the cross-sectional area of the tank and $h(t)$ is the level of the liquid to be controlled.

The output flow through the discharge valve is related to the level of liquid in the tank by (2):

$$q_o(t) = a_v C_v \sqrt{2gh(t)} \tag{2}$$

a_v is the cross-sectional area of the valve orifice and C_v is the discharge coefficient of the valve.

Combining (1) and (2) gives:

$$\dot{h}(t) + a_v C_v \sqrt{2gh(t)} = q_i(t) \tag{3}$$

The input flow is related to the voltage applied to the pump in a linear manner, by means of (4):

$$q_i(t) = K_b v_i(t) \tag{4}$$

K_b is the gain of the pump and $v_i(t)$ is the input voltage of the system.

Finally, (5) shows a relationship between the input voltage $v_i(t)$ and the liquid level inside the tank $h(t)$.

$$\dot{h}(t) + a_v C_v \sqrt{2gh(t)}/A = (K_b/A)v_i(t) \tag{5}$$

The values of the system parameters are shown in Table 1.

Table 1. System parameters.

Symbol	Description	Value
A	Cross-sectional area of the tank	9350×10^{-6} m^2
a_v	Cross-sectional area of the valve orifice	78.50×10^{-6} m^2
C_v	Discharge coefficient of valve	0.2
h_{max}	Maximum liquid level	0.25 m
v_{max}	Maximum input voltage	10 V
K_b	Pump gain	6.66×10^{-6} m^3/sV
g	Gravity constant	9.8 m/s^2

The model of the process is implemented in Simulink to carry out the analysis of the system, the design and the testing of the controllers (Fig. 2).

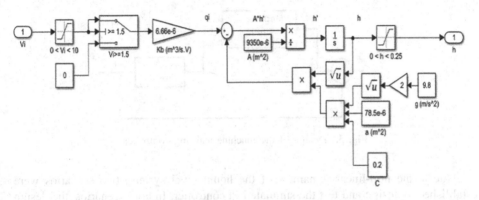

Fig. 2. Process model in Simulink [5].

The validation of the system parameters in this document was not developed, because this work is trying to compare the performance of a controller based on neural networks versus a classic PI design. Therefore, all the analysis is performed on simulated data, using the values specified by the manufacturer in the manual.

2.2 PI Controller Design

The following methodology was used to design the neural controller. First, a PI controller was designed and tested through the simulations (Fig. 3a). Then, the input/output data from the PI was used to train the ANN (Fig. 3b), for this all data were used. Finally, the PI controller was replaced by the machine learning controller based on neural networks (Fig. 3c). Some simulation tests were performed to compare the performance of the non-linear liquid level system when the ANN controller and the PI controller were used.

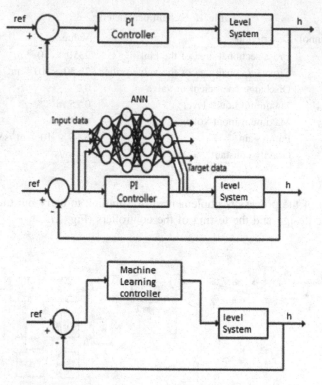

Fig. 3. Design of the machine learning controller.

Due to the non-linear dynamics of the liquid level system, two scenarios were established to design and test the simulated PI controller. In both scenarios, the design criteria was that the controlled system output had the minimum settling time and that the PI controller output did not exceed 10 v, because of this value is the maximum voltage allowed at the pump input.

In scenario 1, a single PI controller was designed so that the non-linear liquid level system worked at the setpoint 0.125 m. In scenario 2, 4 PI controllers were designed in the following setpoints: 0.05, 0.10, 0.15 and 0.20 m. This due to the non-linear dynamics of the plant.

The algorithm that describes the behavior of the PI controller is presented in (6):

$$u(t) = K_p e(t) + K_i \int_0^t e(t) d\tau \tag{6}$$

$u(t)$ is the output signal of the PI controller, which in this case corresponds to the voltage applied to the pump. $e(t)$ is the input signal of the PI controller, which is defined as $e(t) = r(t) - y(t)$, where $r(t)$ is the setpoint and $y(t)$ is the output of the process, that is, the level of liquid in the tank. K_p is the proportional gain and K_i is the integral gain [6].

Applying the Laplace transform to (6), the transfer function of the PI controller is found:

$$U(s)/E(s) = K_p + K_i/s \tag{7}$$

Scenario 1. Because the level of the liquid in the tank can be in the range between 0 and 0.25 m, an intermediate setpoint is chosen $(h = 0.125\,\text{m})$ and a PI controller is designed so that the system carries out an adequate tracking to a step input.

The transfer function of the process, at this setpoint, is:

$$G(s) = \frac{H(s)}{V_i(s)} = \frac{0.0678}{95.12s + 1} \tag{8}$$

Figure 4 shows the response of the system in closed loop when a unit step input is applied. It can be seen that the system has no overshoot, has a settling time $t_s = 372\,\text{s}$, and has a steady state error $e_{ss} = 0.9322$.

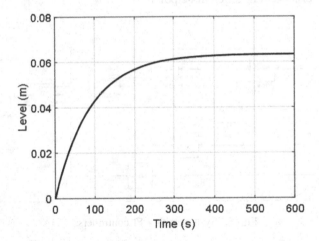

Fig. 4. System response in closed loop.

A PI controller is designed to reduce the settling time and eliminate the steady state error. For the design it is taken into account that the output of the PI controller does not exceed 10 V because this is the maximum voltage supported by the pump. The transfer function of the PI controller is:

$$C(s) = \frac{U(s)}{E(s)} = \frac{79.183s + 4.6718}{s} \tag{9}$$

Scenario 2. Since the plant has a non-linear dynamic, the system is linearized into 4 different setpoints: 0.05, 0.10, 0.15 and 0.20 m. Then, a PI controller is designed for each setpoint, looking for the settling time to be reduced and for the output of the PI controller not to exceed 10 V [7].

The process transfer functions and the PI controllers designed for each setpoint are summarized in Table 2.

Table 2. Process transfer functions and PI controllers.

Setpoint	Linearized plant	PI controller
$h_1 = 0.05\,\text{m}$	$G_1(s) = \frac{0.0429}{60.1587s+1}$	$C_1(s) = \frac{185.2834s+8.055}{s}$
$h_2 = 0.10\,\text{m}$	$G_2(s) = \frac{0.0606}{85.0784+1}$	$C_2(s) = \frac{92.3302s+3.1838}{s}$
$h_3 = 0.15\,\text{m}$	$G_3(s) = \frac{0.0742}{104.1981s+1}$	$C_3(s) = \frac{60.6837s+1.838}{s}$
$h_4 = 0.20\,\text{m}$	$G_4(s) = \frac{0.0857}{120.3181s+1}$	$C_4(s) = \frac{48.8141+1.3193}{s}$

The 4 PI controllers are placed in an array as shown in Fig. 5. The input signal of the system consists of several steps. This allows evaluating the performance of the non-linear liquid level system at different setpoints.

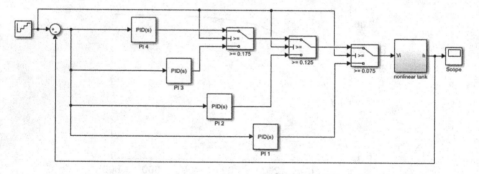

Fig. 5. System with 4 PI controllers.

2.3   ANN Controller

A supervised learning algorithm is part of the machine learning algorithms. It takes a set of input/output data (the training set) and trains a model to generate reasonable predictions for the response to new input data. In addition, supervised learning techniques are formed through classification techniques that predict discrete responses and regression techniques that predict continuous responses.

A machine learning algorithm is a trial and error process [8]. Therefore, it should be taken into account: The training speed, the use of memory, the precision of prediction to new data and finally the transparency or interpretability.

An ANN is part of the supervised learning algorithms, and can be used to model highly non-linear systems. The network is trained iteratively modifying the strengths of the connections so that the inputs are assigned to the correct response.

In this stage several networks with different architectures were trained. These architectures were implemented with a trial and error strategy. Finally, an ANN design with three hidden layers and 20 neurons per layer was selected. Activation functions type TANSIG were used in each layer [9].

The type of network chosen for the design was Feed-Forward Backpropagation. The Levenberg-Marquardt optimization was used for the training function. The gradient descending method was used for the adaptation of the learning function.

3 Results and Discussions

The simulation was conducted towards two scenarios to compare the behavior of the neural controller in contrast to the PI controller. The two scenarios that were made are presented below.

3.1 Scenario 1: Step Input

A supervised learning algorithm is part of the machine learning algorithms. It takes a set of input/output data (the training set) and trains a model to generate reasonable predictions for the response to new input data. In addition, supervised learning techniques are formed through classification procedures that predict discrete responses and regression methods that predict continuous responses.

In order to implement an ANN controller, a PI controller is first designed using Matlab. For this design it was decided that the following parameters were reached: an overshoot of less than 10%, the shortest possible settling time considering that the output signal of the PI controller does not exceed 10 v, which is the maximum voltage that can be applied to the pump.

Based on this PI controller, several neural networks were trained by taking the input data with a step of 0.125 m and the compensator output for the level plant. Several simulation tests were made where the number of layers and neurons per layer was modified. As the system present a high degree of complexity, an ANN with 3 layers and 20 neurons per layer was chosen.

The simulation was developed in the Matlab/Simulink environment. In order to obtain a detailed results analysis of the comparative study between the PI controller and the ANN controller, the RMSE variations were used. The simulation was developed with a time of 200 s.

It can be seen in Fig. 6, that the system response, when the ANN controller is used, has better behavior in transient response compared to the PI, since it does not have an overshoot. In addition, the settling time is much less than 100 s, compared to the case with the PI controller. The control with ANN has a good tracking to the setpoint since there is no steady state error.

In order to obtain a detailed results analysis of the comparative study between the PI controller and the ANN controller, the RMSE variations were used. As can be seen

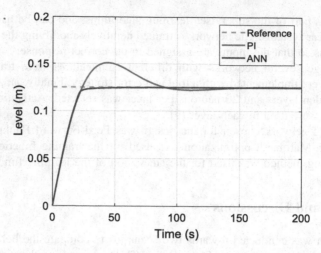

Fig. 6. System response with PI and ANN controller in scenario 1.

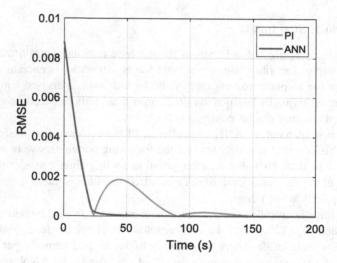

Fig. 7. Comparison of the RSME variation for PI and ANN controllers.

in Fig. 7, the quality of the response is much better with the ANN controller than with the PI controller, therefore the expected results are ratified.

3.2   Scenario 2: Complex Step Input

This scenario was proposed in order to verify the operation of the controller in a more complex environment.

In order to design this ANN, the same steps performed in scenario 1 were executed. The training of the neural network was carried out with the following setpoints: 0.05, 0.10, 0.15 and 0.20 m. However, in order to check the ANN learning the simulation test was performed with 0.04, 0.07, 0.14 and 0.18 m. Many simulation tests were carried out, then the results obtained will be shown.

The ANN of 3 layers and 20 neurons per layer was also chosen in this case. The simulation was developed with a time of 800 s. It can be seen in Fig. 8, that the system controlled with ANN continues having a faster response compared to the one controlled with PI. It can also be noted that the control with PI has a lower steady-state error. This is due to the intrinsic integral action that the PI controller has. On the other hand, the ANN can be expected to learn this type of behavior over time to achieve more accurate results. However, the results obtained are very good. In addition, overshoot is greater for the PI controller compared to the neuronal controller, thus demonstrating good tracking to changes in the setpoint.

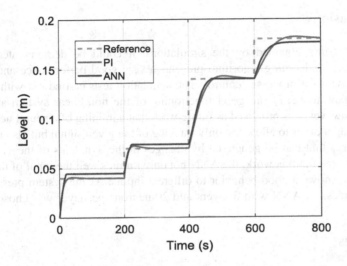

Fig. 8. System response with PI and ANN with 3 layers and 20 neurons per layer controller in scenario 2.

Table 3 shows the RMSE variations for both the PI controller and the ANN controller when the two simulation scenarios are applied. The implemented ANN controller presents a better behavior in the two scenarios compared to the PI controller.

Table 3. RMSE variations for PI and ANN controllers.

Scenario	PI controller	ANN controller
1	0.0253	0.0231
2	0.0089	0.0088

Although the computational cost of the implementation of the neural controller is higher, the system response is noticeably faster than when the PI controller is used. This can be an advantage in systems where it is necessary to minimize the values of overshoot and settling time. However, for processes where these requirements are not necessary, the PI controller would be more viable because of its simplicity of implementation.

In this document a machine learning algorithm was designed using artificial neural networks in order to explore if these methods were applicable to control problems. Because of this work is considered like a first stage, the analysis was carried out using only the simulated model of the plant. That allows to compare if the dynamics of the system controlled with neural networks presents some improvement over the one controlled using a PI. Furthermore, the objective was to realize an analysis of the system response for different setpoints and no analysis of the uncertainties of the process to see how the controllers made an effective rejection of the disturbances.

4   Conclusions

Due to the results obtained by the simulations, it could be demonstrated that the proposed machine learning algorithm presents a very good performance and that it is possible to use it for processes control. The simulation tests carried out with the ANN controller allow to verify the good functioning of the non-linear system at different setpoints. However, it is proposed as future work that a training of a neural network can be performed, in order to allow not only tracking of the given signal but also a rejection of disturbances that can be generated by changes in the dynamics of the system.

As can be seen in this work, the ANN not only mimics well the input of the training data, but also shows a good behavior to different inputs. As the system present a non-linear dynamics, an ANN with 3 layers and 20 neurons per layer was chosen.

References

1. Shariff, R., Cudrak, A., Zhang, Q., Stanley, S.J.: Advanced process control techniques for water treatment using artificial neural networks. J. Environ. Eng. Sci. 3(S1), S61–S67 (2004)
2. Naman, A.T., Abdulmuin, M.Z., Arof, H.: Development and application of a gradient descent method in adaptive model reference fuzzy control. In: TENCON 2000 Proceedings, vol. 3, pp. 358–363. IEEE (2000)
3. Xiao, Y., Hu, H., Jiang, H., Zhou, J., Yang, Q.: A adaptive control based neural network for liquid level of molten steel smelting noncrystlloid flimsy alloy line. In: Proceedings of 4th World Congress on Intelligent Control and Automation, China (2002)
4. Hagan, M.T., Demuth, H.B., Jesús, O.D.: An introduction to the use of neural networks in control systems. Int. J. Robust Nonlinear Control 12(11), 959–985 (2002)
5. Hwang, J.N., Lay, S.R., Maechler, M., Martin, R.D., Schimert, J.: Regression modeling in back-propagation and projection pursuit learning. IEEE Trans. Neural Netw. 5(3), 342–353 (1994)
6. Denuth, H., Beale, M.: Neural Network Toolbox User's Guide for Use with MATLAB. The Math Works Inc., Marde (1996)
7. Xu, J., Ho, D.W., Zheng, Y.: A constructive algorithm for feedforward neural networks. In: 5th Asian Control Conference 2004, vol. 1, pp. 659–664. IEEE (2004)

8. Prasad, V., Bequette, B.W.: Nonlinear system identification and model reduction using artificial neural networks. Comput. Chem. Eng. **27**(12), 1741–1754 (2003)
9. Arel, I., Rose, D.C., Karnowski, T.P.: Deep machine learning-a new frontier in artificial intelligence research [research frontier]. IEEE Comput. Intell. Mag. **5**(4), 13–18 (2010)

Filter Banks as Proposal in Electrical Motors Fault Discrimination

Jhonattan Bulla[1], Alvaro David Orjuela-Cañón[1(✉)] [iD],
and Oscar D. Flórez[2]

[1] Universidad Antonio Nariño, Bogotá D.C., Colombia
{jhonattanbulla, alvorjuela}@uan.edu.co
[2] Universidad Distrital Francisco José de Caldas, Bogotá, Colombia
odflorez@udistrital.edu.co

Abstract. Studies related with the induction motor bearings fault detection have been used digital signal processing and pattern recognition techniques. However, performance of these approaches depends on the use of correct features. This paper deals an analysis of the use of filter banks with uniform and nonuniform frequency subbands to features extraction from vibration signals. Discrimination was developed by an artificial neural network with feedforward connections. Results identifies that the employment of filter banks improve the accuracy in 23% for six considered classes related with faults in bearings.

Keywords: Induction motor · Bearing faults · Filter bank
Artificial neural networks · Feature extraction

1 Introduction

Inside the production line, induction motors (IM) are attractive to be employed because the easier integration capacity and high performance, contrasted with other machines. In this kind of equipment the rolling element bearings are widely used, and their failure is one of the most frequent causes for disruption in manufacture processes [1]. This has inspired a broad researching related with diagnosis techniques based on signal processing to find defects that can affect the productivity of the companies. Examples of this can be shown in different analysis achieved over vibration signals to continuously improve the currently methods [2].

A vibration signal holds information from the fault, load distribution, vibrations induced by the bearing and machinery, and noise. Taking advantage of the signal processing techniques, it is possible to find what component can generate the fault in an IM [3]. This is important in diagnostic processes, where is necessary to extract the information from the data represented by the signals, and in an additional stage, identify the fault.

In this pattern recognition task, the feature extraction step is pertinent, according with the utility in the determination of type of failure and other details as size, location and time. For this reason, currently have been a center of attention in studies related with bearings for IM [4, 5]. In addition, an adequate process of extraction support the classification or identification the faults, improving the results in diagnosis procedures.

© Springer Nature Switzerland AG 2018
A. D. Orjuela-Cañón et al. (Eds.): ColCACI 2018, CCIS 833, pp. 50–62, 2019.
https://doi.org/10.1007/978-3-030-03023-0_5

Digital signal processing techniques have been useful for different problems with focus in feature extraction for classification tasks. In the beginning, the signal processing techniques were mainly based on statistical parameters to obtain representative information. Features as RMS, mean, kurtosis and crest factor values have been widely used due to its sensitive to the shape of the signal [3]. These time-domain techniques have been commonly employed in the industry for detect faults in induction motors [6]. Alternatively, the frequency-domain methods have provided another approach to detect the faults in IM. Fast Fourier transform (FFT) is a common method to produce a spectrum and to analyze the signal in its frequency components. Also, strategies related with the spectrum analysis are based on power spectral density functions, Cepstrum analysis, narrow band envelope analysis and others [3, 7]. Finally, a domain composed by time and frequency simultaneously, analyzes non-stationary signals, which can represent problems for techniques as FFT [7, 8].

Filter banks are approaches used to develop multirate systems in communications, however applications can be seen in the industrial mode for obtain useful features. Two main proposals of these banks are given by uniform and nonuniform subbands, according with its regular or irregular division of the all frequency band available. The last one can be implemented in a similar way as analysis with the Wavelet transform and is the most employed [9, 10]. Applications of filter banks in specific case for faults in induction motors can be seen in [11–14].

The computation of these parameters has allowed the employment of several tools such as artificial neural networks (ANN), support vector machines (SVM), discriminant analysis, fuzzy logic and others, due to this job is highly complex and time demanding [7]. These techniques from computational intelligence have been increased to automate the decision making process. ANN is a method usually used to develop this task. Examples of this can be seen in [15], where the employment of time and frequency domains trained a multilayer perceptron (MLP). Using feature extraction through wavelet transform analysis and computation of statistical measures, it was shown that this machine learning proposal can be useful in fault diagnosis [15]. Also, a detailed work based on self-organizing maps and principal component analysis was implemented an optimal classification for faults in three phase induction motors [16]. Neuro-fuzzy inference was employed as classification, using multi-scale entropy and wavelet decomposition for feature extraction for diagnosis in rotatory machinery in [12, 17]. SVM was implemented for bearing fault detection with the use of wavelet decomposition to extract features, compared with a classification developed by an ANN, The results were comparable, but the easily implementation of SVM was notable [13].

Present work show approximations from the use of filter banks with uniform and nonuniform subbands for feature extraction. Making use of the same filter design, two filter banks were proposed to get characteristics in the signal to determine the classification of faults for IM. In spite of previous studies, this one has as focus the analysis of techniques as time, frequency or time-frequency domains as explained in [18]. A similar study was carried out for analysis of vibration signals but with high-resolution spectral analysis [14].

2 Methodology

Two approaches were employed for our comparison: one of them was based on feature extraction from signals in time domain, and the second one was based on parameters obtained from signals in frequency domain. Both approaches are detailed in this section with aspects of database and the implementation for the fault discrimination.

2.1 Database

Signals in the used database for this study was provided by the Case Western Reserve University Bearing Data Center [19]. Artificial fissures with diameter of 0.007 and depth of 0.011 in., and motor load from 0 to 3 hp were included. The shaft speed was in the interval 1230 to 1797 rpm and model number SKF 6205-2RS JEM SKF.

According with the defective bearing location was determined the type of defect in the IM. Three main positions were identified: rolling element, inner and outer races. This last location was divided into three categories according to the fault position relative to the load zone: 'centered' (fault in the 6.00 o'clock position), 'orthogonal' (3.00 o'clock) and 'opposite' (12.00 o'clock). In total, these six points established the classes to obtain for the present study.

2.2 Preprocessing

First, a filter with cutoff frequency at 10 kHz was used to warranty that the relevant information was included in the features. An experimental way allowed observing that the after 10 kHz just a noise component was present in the signal.

According with the analysis developed in [15], all frequencies of the components included in the ball bearing were computed, and then, it was determined the minimum frequency of 2549 Hz corresponding by the outer race. In this way, signal was divided into nonoverlapping segments of 2048 samples. This number was chosen for the representation of the signal due that the minimum was the 1665 samples. This measure warranties the at least a cycle of the normal condition and the different frequencies obtained in its components. For all considered faults 200 segments were employed for classification.

2.3 Filter Banks

Each subband was processed in an independent way to extract the features. Complete frequency band of input signal $x[n]$, given by the segmentation in the preprocessing was divided into four subbands. For this, a first approach took advantage of the dyadic analysis filter bank (DAFB), where a tree structure of two levels was employed based on a basic unit composed by a low-pass (LP) and high-pass (HP) filters followed by a decimation process with a factor of 2. LP and HP are halfband filters with a cutoff frequency of fs/4 (see Fig. 1a) [9].

Spectrum of vibration signals is certainly nonuniform over the frequency range, so its subbands can have different characteristics. This kind of analyses makes possible to process each subband with different strategies. Then, a second approach modified the

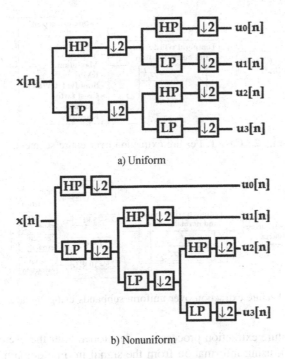

a) Uniform

b) Nonuniform

Fig. 1. Dyadic structures for the filter banks

tree structure in an asymmetric way, based on three levels but with the same number of subbands in the output of the bank (see Fig. 1b). Both proposals of filter banks used the same basic unit, being designed just one filter employing the Parks McClellan algorithm, with transition band wide of 500 Hz and attenuation of −70 dB in the stop band. The filter design has not considerations for perfect reconstruction [9, 10].

Each input signal $x[n]$ (Fig. 1) was presented to the filter bank, obtaining $u_m[n]$ filtered signals, where m is the number of the subbands, being four for our study. For each one of these $u_m[n]$ signals were computed the same features with the aim to compare the use of both filter banks in terms of classification. Next subsection describes each parameter computed.

2.4   Feature Computation

Before the feature computation, a normalization process was developed, where mean and standard deviation values were used to normalize the signal.

Five approaches were established to compare the use of filter banks:

Case 1: the feature extraction was implemented in the segment of study established and mentioned in the preprocessing subsection (see Fig. 2), using information from the signal in time domain.

Case 2: the feature extraction was implemented after a filtering process with a uniform filter bank (Fig. 3), using information from the signal in time domain.

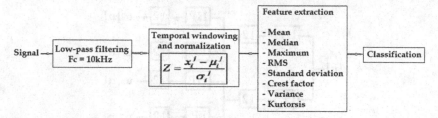

Fig. 2. Case 1: Feature extraction over entire segment.

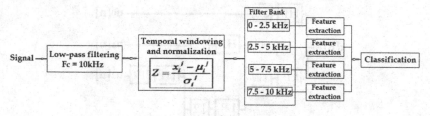

Fig. 3. Case 2: Feature extraction over uniform subbands using signal in time domain.

Case 3: the feature extraction process was developed after the use of a nonuniform filter bank (Fig. 4), using information from the signal in time domain.

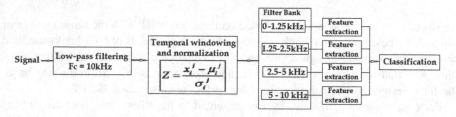

Fig. 4. Case 3: Feature extraction over nonuniform subbands using signal in time domain.

Case 4: the feature extraction process was developed after the use of a nonuniform filter bank (Fig. 5), using information extracted from signal in frequency domain (Fast Fourier - FFT box).

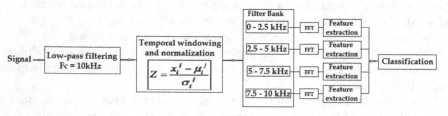

Fig. 5. Case 4: Feature extraction over nonuniform subbands using signal in frequency domain.

Case 5: the feature extraction process was developed after the use of a uniform filter bank (Fig. 6), using information extracted from signal in frequency domain (FFT box).

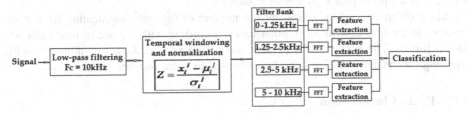

Fig. 6. Case 5: Feature extraction over uniform subbands using signal in frequency domain.

In each case, eight parameters were computed in the signal in time and frequency domains. For cases two and three the calculation was implemented for every signal $u_m[n]$ and for cases four and five were computed in the magnitude of signal in frequency domain. Statistical parameters were considered according with previous works [13, 15, 17, 20], where the mean (1), median (2) and maximum (3) values were considered. Also, variance (4) and standard deviation (5) were included due to information about the dispersion of signal around their reference mean value. Root mean square (RMS) value (6) increases gradually as fault happened, in this way, provide information of incipient fault stage while it increases with the fault development [21]. Kurtosis (7) quantifies the peak value of the probability density function and for normal rolling element bearing has a value of 3. Finally, Crest factor (CF) value (8) computes the impact occurred in the rolling element and the race contacts [21].

Expressions for eight considered parameters can be obtained in the way:

$$\bar{x} = \frac{\sum_{i=1}^{n} x_i}{n} \tag{1}$$

$$Median = \begin{cases} x_{n/2} & if\ n\ is\ even \\ \frac{x_{n/2} + x_{(n+1)/2}}{2} & if\ n\ is\ odd \end{cases} \tag{2}$$

$$x_{max} = \max(x) \tag{3}$$

$$RMS = \sqrt{\frac{\sum_{i=1}^{n} (x_i - \bar{x})}{n}} \tag{4}$$

$$\sigma^2 = \frac{\sum_{i=1}^{n} (x_i - \bar{x})^2}{n - 1} \tag{5}$$

$$\sigma = \sqrt{\sigma^2} \tag{6}$$

$$K = \frac{\sum_{i=1}^{n} (x_i - \bar{x})^4}{(n - 1)\sigma^4} \tag{7}$$

$$CF = \frac{x_{max}}{RMS} \qquad (8)$$

where x is the processed signal with n samples.

Main differences of the cases are the number of features, augmenting from eight features in the first case to 32 in other cases, according with the use of four subbands for each case. The scope of the present work is to present these preliminary results, showing this analysis, which can be augmented to more number of subbands.

2.5 Faults Classification

MLP architecture was composed by three layers, which have been adequate to solve most of the classification problems [22, 23]. Inputs were formed by the features computed as explained in the previous section. Output layer was composed by six nodes, each one for each class to be determined. In the hidden layer a heuristic methodology was conducted, testing from two to then nodes. This allows to evaluate how is the relation between the number of nodes and the faults classification problem in a trial-error process. Also, currently is an open question with multiple solutions but without a consented [24]. Hyperbolic tangent was used as activation function in the hidden layer and a softmax function was used in the output layer according with the properties that a logistic function has [25].

For each proposed architecture (number of nodes in the hidden layer), 100 different initializations were implemented to determine the robustness to initial conditions in the synaptic weights, and in this way, estimating what happen with the local minima.

Cross-validation technique was employed to validate the generalization of the network training. Four folds were applied with 50 segments in each subset. Six classes were analyzed according with the location of the fault: normal, inner race, rolling element (ball), outer race centered, outer race orthogonal and outer race opposite. Classification criterion was given when just one of the nodes in the output layer was activated (+1). If two or more nodes were activated the input was not classified.

3   Results

Tables 1, 2, 3, 4 and 5 show the results for the five cases taken into account. The tables show the best result obtained from the 100 initializations and for each subset of the cross-validation technique and the mean for the four folds.

In the case 1, best result was reached with an accuracy of 80.33% as maximum with eight nodes in the hidden layer. Mean of value for four subsets was 76.25% with the same result for the architecture with nine nodes in the hidden layer. Table 6 shows the confusion matrix for the MLP with the best result.

An accuracy of 100% was obtained in the second case. In this opportunity an ANN with three nodes in the hidden layer was enough to classify the input in a precise outline. Best mean for cross-validation was achieved with the same number of nodes in the hidden layer. Same result was produced with five nodes, but a simpler model

Table 1. Best results for the Case 1.

Nodes in hidden layer	Folds				Mean
	1	2	3	4	
2	58.00	62.00	57.67	64.67	60.58
3	74.00	72.33	72.33	77.33	74.00
4	73.33	73.67	72.00	78.33	74.33
5	74.67	76.00	71.33	79.00	75.25
6	75.67	76.00	71.67	79.33	75.67
7	75.67	76.33	72.00	79.33	75.83
8	**76.00**	76.33	72.33	**80.33**	**76.25**
9	75.67	**77.33**	**72.67**	79.33	76.25
10	76.00	76.33	72.00	79.00	75.83

Table 2. Best results for the Case 2.

Nodes in hidden layer	Folds				Mean
	1	2	3	4	
2	91.67	91.33	92.33	92.67	92.00
3	99.00	**100.00**	**99.33**	99.33	**99.41**
4	99.33	99.67	99.00	99.33	99.33
5	**99.67**	99.67	99.00	99.33	99.41
6	99.00	99.33	99.33	99.00	99.16
7	98.67	99.33	99.00	**99.67**	99.16
8	99.33	99.33	99.33	99.00	99.24
9	99.33	99.33	98.33	98.67	98.91
10	99.33	98.67	98.67	99.67	99.08

Table 3. Best results for the Case 3.

Nodes in hidden layer	Folds				Mean
	1	2	3	4	
2	83.67	87.33	85.33	86.67	85.75
3	96.00	**98.33**	**98.67**	97.67	**97.66**
4	**97.00**	97.33	98.00	97.67	97.50
5	96.00	97.67	98.33	97.33	97.33
6	96.33	97.33	98.67	97.33	97.41
7	95.33	97.33	97.67	97.33	96.91
8	96.33	97.67	98.00	**98.33**	97.58
9	95.33	97.00	98.00	96.33	96.66
10	95.33	97.67	97.67	97.00	96.91

Table 4. Best results for the Case 4.

Nodes in hidden layer	Folds				Mean
	1	2	3	4	
2	99.00	98.00	97.00	97.67	97.92
3	**100.00**	**100.00**	**100.00**	**100.00**	**100.00**
4	100.00	100.00	100.00	100.00	100.00
5	100.00	100.00	100.00	100.00	100.00
6	100.00	100.00	100.00	100.00	100.00
7	100.00	100.00	100.00	100.00	100.00
8	100.00	100.00	100.00	100.00	100.00
9	100.00	100.00	100.00	100.00	100.00
10	100.00	100.00	100.00	100.00	100.00

Table 5. Best results for the Case 5.

Nodes in hidden layer	Folds				Mean
	1	2	3	4	
2	99.00	95.67	97.33	99.00	97.75
3	**100.00**	**100.00**	99.67	**100.00**	**99.92**
4	100.00	100.00	99.67	100.00	99.92
5	100.00	100.00	99.67	100.00	99.92
6	100.00	100.00	99.67	100.00	99.92
7	100.00	100.00	99.67	100.00	99.92
8	100.00	99.67	**100.00**	100.00	99.92
9	100.00	100.00	99.67	100.00	99.92
10	100.00	100.00	99.67	100.00	99.92

Table 6. Confusion matrix with the best result for the Case 1

Classes	1	2	3	4	5	6
Normal	**45**	0	0	0	0	0
Inner race	0	**42**	0	0	0	7
Rolling element (ball)	5	0	**35**	0	0	1
Outer race (centered)	0	7	0	**48**	0	4
Outer race (orthogonal)	0	0	3	0	**45**	11
Outer race (opposite)	0	1	2	0	3	**26**
No class	0	0	0	0	2	1

always is preferred. Table 7 shows the confusion matrix for this case, where it is possible to see the diagonal with the number of segments used in the validation.

For third case, best result for accuracy was 98.67% with three nodes in the hidden layer of the ANN. For the result of the classification with the four folds, an accuracy of

Table 7. Confusion matrix with the best result for the Case 2

Classes	1	2	3	4	5	6
Normal	**50**	0	0	0	0	0
Inner race	0	**50**	0	0	0	0
Rolling element (ball)	0	0	**50**	0	0	0
Outer race (centered)	0	0	0	**50**	0	0
Outer race (orthogonal)	0	0	0	0	**50**	0
Outer race (opposite)	0	0	0	0	0	**50**
No class	0	0	0	0	0	0

Table 8. Confusion matrix with the best result for the Case 3

Classes	1	2	3	4	5	6
Normal	**50**	0	0	0	0	0
Inner race	0	**49**	0	0	0	0
Rolling element (ball)	0	0	**48**	0	0	1
Outer race (centered)	0	0	0	**50**	0	0
Outer race (orthogonal)	0	0	1	0	**50**	0
Outer race (opposite)	0	1	1	0	0	**49**
No class	0	0	0	0	0	0

97.66% was reached with a model with three nodes in the hidden layer, too. Table 8 shows the results in a confusion matrix with the missclassifications.

Forth case is shown in Table 4, where a classification rate of 100% was reached with two nodes in the hidden layer for the mean of all folds. A confusion matrix is not presented because is identical as Table 7 with all classes well classified as shown in the diagonal of the matrix.

Finally, the fifth case is resumed in Table 5 with the best classification rate of 100% and 99.92% in mean for all folds. Also, two nodes were enough to discriminate the faults. Confusion matrixes are shown for cases one to three (Tables 6, 7 and 8), cases four and five had tables as case 2 (see Table 7).

4   Discussion

Results could visualize the enhanced in terms of classification when a processing employing filter banks was developed. Cases two and three are evidence of this, where the accuracy was improved from 76% to 99% in a general way. It is important to note that this is an unfair comparison due to the number of inputs for the ANN, being eight in the first case contrasted with 32 features in the last mentioned case. In addition, the nodes in the hidden layer was reduced from eight in the first case to just three in last both cases. Nevertheless, filter banks approximation can be a helpful proposal for diagnostic systems implementation based on multirate systems, which can be developed to work in an online manner offering promissory results. When information was extracted from the signal in the frequency domain, results reach the best results with models with few nodes in the hidden layer (see Tables 4 and 5).

Differences between the use of uniform and nonuniform subbands were found for time and frequency approaches. For time domain, this distinction represents a 1.75% of improvement in the accuracy when uniform bands were applied to extract features, reaching the 100% in one of the simulations. Frequency domain exhibited and improvement of 0.08% with the use of nonuniform bands, maintaining the same number of nodes in the hidden layer.

When confusion matrixes from cases one and three are compared (see Tables 6 and 7), it is possible to observe that the classes related with faults in the outer race centered and orthogonal, and the normal class have the best results. Misclassifications are found in the other three classes, noticing that the most difficult fault to classify was the defect in the outer race opposite with a 52% of its classification. Likewise, the fault in the rolling element is a laborious deficiency to find, with a detection rate of 70% for the case 1.

Comparing the present result with previous studies, similar results were found in [15], where the outer race was the fault more difficult to classify with a 93.33% when features extracted from time-domain were used. Also, results are comparable with [26], where just features based on Wavelet criteria reached accuracies between 62.3% and 99.68%. Employing statistical parameters and ANN with six classes, results between 93% and 98% in accuracy were reached in [27]. In a general way, results are in the interval of values found in the literature, but with different conditions related with the feature extraction and specific classes used for the classification.

5 Conclusions

According with these preliminary results, feature extraction methods based on implementation of filter banks have been presented, exploring alternatives from digital signal processing area. Results showed that filter banks with uniform subbands, in spite of augmented of number of features represent an improvement of the accuracy in terms of classification of six faults considered for the present study. Also, when features were obtained in the frequency domain the results show a considerable improvement with models with just two units in the hidden layer. Finally, the present analysis opens the door to the use of these techniques that can be implemented to operate in online diagnosis systems.

Acknowledgment. Authors thank the Universidad Antonio Nariño under project 2017211 and code PI/UAN-2018-628GIBIO. Also, Universidad Distrital Francisco Jose de Caldas contributed with the support and financial assistance in this work.

References

1. Randall, R.B., Antoni, J.: Rolling element bearing diagnostics-a tutorial. Mech. Syst. Sig. Process. **25**, 485–520 (2011)
2. Smith, W.A., Randall, R.B.: Rolling element bearing diagnostics using the case western reserve university data: a benchmark study. Mech. Syst. Sig. Process. **64**, 100–131 (2015)
3. El-Thalji, I., Jantunen, E.: A summary of fault modelling and predictive health monitoring of rolling element bearings. Mech. Syst. Sig. Process. **60**, 252–272 (2015)
4. Caesarendra, W., Tjahjowidodo, T.: A review of feature extraction methods in vibration-based condition monitoring and its application for degradation trend estimation of low-speed slew bearing. Machines **5**, 21 (2017)
5. Wang, H., Chen, P.: A feature extraction method based on information theory for fault diagnosis of reciprocating machinery. Sensors **9**, 2415–2436 (2009)
6. Chebil, J., Hrairi, M., Abushikhah, N.: Signal analysis of vibration measurements for condition monitoring of bearings. Aust. J. Basic Appl. Sci. **5**, 70–78 (2011)
7. Rai, A., Upadhyay, S.H.: A review on signal processing techniques utilized in the fault diagnosis of rolling element bearings. Tribol. Int. **96**, 289–306 (2016)
8. Liu, J., Wang, W., Golnaraghi, F., Liu, K.: Wavelet spectrum analysis for bearing fault diagnostics. Meas. Sci. Technol. **19**, 15105 (2007)
9. Porat, B.: A Course in Digital Signal Processing. Wiley, New York (1997)
10. Strang, G., Nguyen, T.: Wavelets and Filter Banks. SIAM, Philadelphia (1996)
11. Rafiee, J., Rafiee, M.A., Tse, P.W.: Application of mother wavelet functions for automatic gear and bearing fault diagnosis. Expert Syst. Appl. **37**, 4568–4579 (2010)
12. Lou, X., Loparo, K.A.: Bearing fault diagnosis based on wavelet transform and fuzzy inference. Mech. Syst. Sig. Process. **18**, 1077–1095 (2004)
13. Konar, P., Chattopadhyay, P.: Bearing fault detection of induction motor using wavelet and Support Vector Machines (SVMs). Appl. Soft Comput. **11**, 4203–4211 (2011)
14. Garcia-Perez, A., de Jesus Romero-Troncoso, R., Cabal-Yepez, E., Osornio-Rios, R.A.: The application of high-resolution spectral analysis for identifying multiple combined faults in induction motors. IEEE Trans. Ind. Electron. **58**, 2002–2010 (2011)
15. Zarei, J.: Induction motors bearing fault detection using pattern recognition techniques. Expert Syst. Appl. **39**, 68–73 (2012)
16. Ghate, V.N., Dudul, S.V.: Optimal MLP neural network classifier for fault detection of three phase induction motor. Expert Syst. Appl. **37**, 3468–3481 (2010)
17. Zhang, L., Xiong, G., Liu, H., Zou, H., Guo, W.: Bearing fault diagnosis using multi-scale entropy and adaptive neuro-fuzzy inference. Expert Syst. Appl. **37**, 6077–6085 (2010)
18. Kia, S.H., Henao, H., Capolino, G.-A.: Some digital signal processing techniques for induction machines diagnosis. In: 2011 IEEE International Symposium on Diagnostics for Electric Machines, Power Electronics & Drives (SDEMPED), pp. 322–329 (2011)
19. Loparo, K.A.: Case Western Reserve University Bearing Data Center (2012)
20. Wang, D., Peter, W.T., Tsui, K.L.: An enhanced Kurtogram method for fault diagnosis of rolling element bearings. Mech. Syst. Sig. Process. **35**, 176–199 (2013)
21. Yiakopoulos, C.T., Gryllias, K.C., Antoniadis, I.A.: Rolling element bearing fault detection in industrial environments based on a K-means clustering approach. Expert Syst. Appl. **38**, 2888–2911 (2011)
22. Haykin, S.: Neural Networks and Learning Machines. Prentice Hall, Englewood Cliffs (2009)
23. Hornik, K.: Approximation capabilities of multilayer feedforward networks. Neural Netw. **4**, 251–257 (1991)

24. Sheela, K.G., Deepa, S.N.: Review on methods to fix number of hidden neurons in neural networks. Math. Probl. Eng. **2013**, 7–9 (2013)
25. Bishop, C.: Pattern Recognition and Machine Learning. Information Science and Statistics, 1st edn. Springer, New York (2007). Corr. 2nd printing edn. Springer, New York (2007)
26. Zaeri, R., Ghanbarzadeh, A., Attaran, B., Moradi, S.: Artificial neural network based fault diagnostics of rolling element bearings using continuous wavelet transform. In: 2011 2nd International Conference on Control, Instrumentation and Automation (ICCIA), pp. 753–758 (2011)
27. Prieto, M.D., Cirrincione, G., Espinosa, A.G., Ortega, J.A., Henao, H.: Bearing fault detection by a novel condition-monitoring scheme based on statistical-time features and neural networks. IEEE Trans. Ind. Electron. **60**, 3398–3407 (2013)

Discrimination of Nonlinear Loads in Electric Energy Generation Systems Using Harmonic Information

Juan de Dios Fuentes Velandia[1], Alvaro David Orjuela-Cañón[2](✉) ⓘ, and Héctor Iván Tangarife Escobar[1]

[1] Servicio Nacional de Aprendizaje, Bogotá D.C., Colombia
[2] Universidad Antonio Nariño, Bogotá D.C., Colombia
alvorjuela@uan.edu.co

Abstract. This paper contains a proposal to determine the kind of nonlinear load when are connected to the solar or conventional generation system. A database was built with sampled signals extracted from the photovoltaic system of the National Learning Service (SENA) in Bogota, Colombia. The used methodology has an acquisition system of voltage signals, and then, information from harmonic distortion was employed to identify the nonlinear loads. An artificial neural network was implemented to discriminate appliances with supervised learning. Two proposals were implemented. First one was based on energy information and second one was worked with wave peaks information. Results show that a classification rate of 95% could be reached in a problem with eight classes.

Keywords: Power quality · Signal processing · Neural networks
Electric energy generation systems

1 Introduction

Electric power generation from photovoltaic systems has been implemented around the world in an incrementally manner. This representative generation way has been presented as a new, clean and ecological mode to obtain electric energy in the planet, according with the decreasing of fossil combustibles use [1]. Photovoltaic systems (PV) are based on solar panels and have the capacity to convert the irradiance from the sun through direct current (DC) power. This kind of power is transformed to alternating current (AC) using a power inverter. Then, this device is connected to the public service grid to be used in homes and industry [2] (see Fig. 1). This power generation depends on number of hours that the solar energy is available through the irradiance. Also, aspects as type of panel, quantity and mode of connections between them, orientation, inclination and general quality of installation make that use of this resource will be better in some cases than others [3].

In Colombia, different PV systems have been installed taking advantage of geolocation. Examples of this can be seen in the Guajira region at north of the nation, contributing with electric power generation native population [4]. Also, studies related

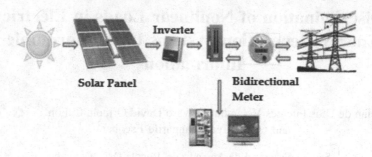

Fig. 1. PV system

with the use this technology, its implementation and development with different government plans have been developed in Medellin city [5]. In a general way, some works have been collected, which are explained from the distributed generation point of view [6].

Unfortunately, progresses in this new power generation alternatives have new challenges. Employed electronics circuits in the conversion stage affects the power quality (PQ) of the system, producing harmonic distortion (HD) [7]. At the same time, nonlinear loads connected to the grid also modify the PQ. This happens when the electronics circuits that have as function switching tasks are the main cause of this problem [8, 9]. Standards have been developed to regularize the issues caused by inconveniences created by the HD [10]. Also in Colombia there is a set of norms for mitigating some aspects related with PV systems reported in the norm NTC 4405, indicating some instructions for energy efficiency and its evaluation when these systems are employed.

In spite of these national and international efforts, these topics are relatively new in the Colombian context, making necessary more studies related with the solar resources potential and the implications in terms of PQ and HD. In this way, few studies have been reported in the PQ area for PV systems, but limited with regard to deeper analysis and further details about the implementation and the results [11, 12].

On the other hand, other works have worked the HD problem to determine the cause [13]. There, the voltage waveform was used to discriminate phenomenon as transients and flickers that modify the original wave, modifying the PQ of the system. Also, Raptis in 2015 showed the advantages of artificial neural networks (ANN) compared with fuzzy systems to determine the total PQ, employing different features extracted from voltage signals [14]. More applications of ANN have been used to control active filters, regulating the effects of nonlinear loads that modify the PQ in a system [15]. Finally, reviews of studies where computational intelligence techniques as previous stage after the use of digital signal processing tools have been used in the PQ area, showing some beneficial aspects of this utilization [16, 17].

This work presents the employment of the National Learning Service (SENA, for Spanish initials) PV system in Colombia. This institution developed the project called SUSENA, which was implemented with the German development and economical ministry. As product of this collaboration, it was installed a solar power plant with four sections, divided into 145 modules with a total capacity of 20.45 kW of DC power. This power is connected to the grid, employing monophasic power inverters to generate electrical energy in a building at the south of Bogota city.

Based on the mentioned problems related with nonlinear loads, the purpose of this study was to evaluate how these loads can be identified under two scenarios: when the power generation was obtained from the sun and when the power generation was obtained from the traditional power grid. In this way, the presented results try to give some new tools to analyze PQ based on digital signal processing and computational intelligence stages.

2 Methodology

Four stages were established to carry out the methodology in the present work: conditioning signal, acquisition signal, digital processing and classification. First, it is necessary to build a database based on voltage waveform signals extracted from the PV system. Then, an analysis of the signal in the frequency domain was implemented to obtain the features, which were used in the classification stage. Figure 2 shows the employed stages in the used methodology.

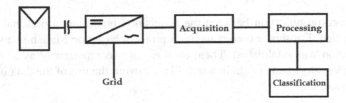

Fig. 2. Implemented methodology.

2.1 Signal Acquisition and Conditioning

For the Colombian electrical grid, frequency is given by 60 Hz for voltage waveform, obtaining a period of 16 ms approximately. Amplitude of the voltage signal reaches 170 v in its peak and a root mean square value of 120 v. These characteristics were taken into account to design the acquisition system.

An Arduino UNO board was employed to implement the analogic-digital conversion. This board holds an Atmega 328 microcontroller with maximum amplitude value limited to 5 v in the input ports, without the possibility of obtaining negative values. Then, it was necessary to use a transformer to reduce the voltage amplitude with a 1:10 relation. After, a DC offset of 2.5 v was adjusted to acquire negative values. Acquisition system was programmed to register data with a sample rate of 1.6 kSamples/s. This established a bandwidth of 833 Hz for this kind of signals, according with Nyquist theorem [18].

Figure 3 shows the implemented circuit, where it is to possible to see the transformer and the tension divider. This last is given by the R1 and R2 resistances, which were placed to reduce the voltage before the input port of the Arduino board. Power supply of these circuits is based on a rectifier system operated over the same tension applied in the transformer input (see Fig. 4).

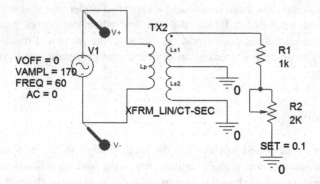

Fig. 3. Voltage reduction.

Addition of signals was developed using an operational amplifier as non-inverter adder in the rail-to-rail configuration, which does not need high voltage values for its polarization.

Finally, a communication between the acquisition board and a personal computer was implemented. For this, a communication protocol based on Matlab software and a serial connection was established. Then, each signal was represented as a vector with 100000 samples, storing this data in a text file, allowing the use of the data in different types of software.

2.2   Employed Nonlinear Loads

As the objective of this work is to discriminate loads connected to the grid, different types of appliances were considered. Linear and nonlinear loads and its combinations were employed. For this, a light emitting diode (LED) lamp, a fluorescent lamp and an induction motor were chosen, according with the implications in terms of HD [18].

Acquisitions without any load also were achieved. For this, 50 signals were obtained for this no-load scenario. Then, for each type of load, 20 acquisitions were accomplished for seven additional classes given by the three mentioned loads. In this way, it was possible to analyze the effect of each load and the combination of them. Table 1 displays the used classes in the study.

After obtaining the text files with the signals represented as vectors, each one of them was divided into segments to augment the elements obtained in the database. For this, each acquisition of 100000 samples (one minute of acquisition) was divided into six segments of 10 s without overlapping. In this way, classes two to eight had sets with 200 elements and the class one had a set with 500 elements. This was performed to have a considerable number of examples to train the classifier.

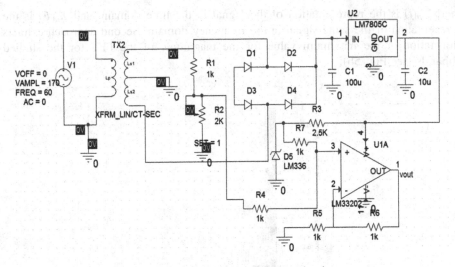

Fig. 4. Acquisition conditioning circuit.

Table 1. Classes used in the study.

Class	Type of load
Class 1	Without load
Class 2	Fluorescent lamp
Class 3	LED lamp
Class 4	Fluorescent lamp + LED lamp
Class 5	Induction motor + fluorescent lamp
Class 6	Induction motor + LED lamp
Class 7	Induction motor
Class 8	Induction motor + fluorescent lamp + LED lamp

2.3 Feature Extraction

After of having the segments from the voltage signals, an analysis was achieved in the frequency domain. For this, in each segment was applied the Fast Fourier Transform (FFT), and then, magnitude was computed. As the available bandwidth corresponds to 833 Hz, this spectrum was divided into 15 subbands of frequency, according with the number of harmonics for the 60 Hz frequency.

Signals of each segment were represented with information from each subband. Two approaches were implemented to obtain vectors with 15 features that denote each segment. First approach computed the power spectral density (PSD) for the subbands (see Fig. 5a), according with expression (1), in the way:

$$E = \int_{-\infty}^{\infty} |x_a(t)|^2 dt = E = \int_{-\infty}^{\infty} |X_a(F)|^2 dF \qquad (1)$$

where $x_a(t)$ is the representation of the signal in the time domain, and $X_a(F)$ is the representation of the same signal in the frequency domain. Second approach extracts information of the maximum value for the magnitude of the FFT for the studied subbands (see Fig. 5b).

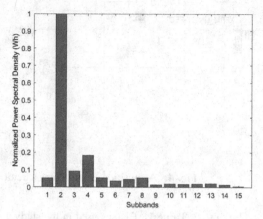

a) PSD information for each subband

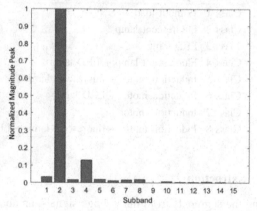

b) Maximum value for the FFT magnitude for each subband

Fig. 5. Example of features for both approaches for a signal ot class 1.

2.4   Model Selection for Discrimination

For discrimination, a model based on ANN was implemented to determine the type of load from the features extracted in the frequency domain analysis. For this, it is necessary to present the information of the 15 characteristics to the neural network (see Fig. 5), and adjust the synaptic weights according with the information of the load in a supervised way [19, 20].

The employed ANN for this task was Multilayer Perceptron (MLP), where synaptic weights have feedforward connection and making an input-output mapping. Number of inputs was 15, according with the number of features obtained in the previous stage. Just one hidden layer was used, corresponding with the universal theorem, which discuss about the use of one layer in different number of applications [19, 20]. For this layer, the number of units or nodes was found in an experimental way, testing between one and ten units. Output layer was determined for the number of classes, with eight units in this problem. All nodes had the hyperbolic tangent function as transfer function in both layers. Finally, the algorithm employed for training was resilient backpropagation [21, 22].

Cross-validation was implemented to analyze the generalization of the trained models [23]. Database was divided into five subsets or folds to visualize the generalization and evaluate the learning of the network. Then, the model was trained with four folds and evaluated with the fold left out training.

For comparison two ways of power generation (PV systems and traditional grid) were studied. From these modes three scenarios for discrimination were implemented: (i) using information from PSD, (ii) using information from magnitude peaks, and (iii) using information from PSD and magnitude peaks.

3 Results

Information was based on voltage signals acquired in the interconnected grid, taking generation from PV system and connection to the traditional grid. All process was carried out in the SENA campus at south of Bogota city.

As an illustrative manner, Fig. 6 shows the results for discrimination when features from PSD and magnitude peaks were employed simultaneously while the PV system supplied the power. Results for the five folds are presented where the generalization performance for the different models is observed. It is possible to see that the number of units in each model and the effect in the results. Boxplots visualize the results given by 100 different initializations for the network parameters. Figure 7 show in a similar way the results for the power generated by the traditional grid.

For each power generation case, comparisons are shown related with the feature extraction scenario. Table 2 shows the results for the three situations when power generation from PV system, reporting the best result from the 100 initializations. Bold numbers represent the highest value for each employed method (PSD, peak values and both). In addition, Table 3 shows the results when power generation was taken from the traditional grid. In the two tables, standard deviation was computed with the best value of each fold.

Finally, Table 4 shows the confusion matrix for the best case when the PV system was employed. It is possible to see how the last four classes had a worse performance than the first classes, where the accuracy reaches a 100% for the classification. Also, it is notable that class five (induction motor with florescent lamp) was the most difficult to discriminate with an accuracy of 67.5%.

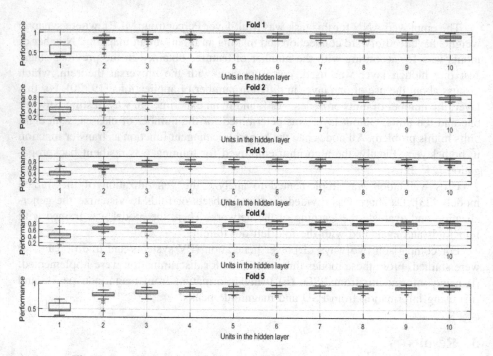

Fig. 6. Results for discrimination with generation from the PV system

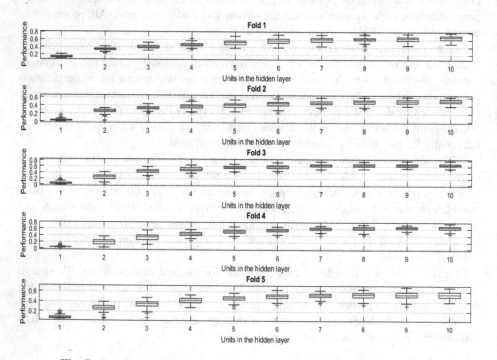

Fig. 7. Results for discrimination with generation from the traditional grid.

Table 2. Results for energy supplied by the PV system.

Units in the hidden layer	PSD	Magnitude peaks	PSD and peaks
1	66.77 + 5.66	86.86 + 1.19	70.00 + 3.24
2	83.27 + 5.37	**94.43 + 4.98**	84.00 + 7.21
3	85.31 + 6.38	76.38 + 7.83	87.81 + 5.73
4	86.25 + 7.06	76.93 + 7.84	89.06 + 5.78
5	87.22 + 6.75	77.50 + 7.54	91.37 + 4.85
6	88.06 + 6.00	78.31 + 8.08	93.28 + 5.04
7	88.06 + 5.74	77.31 + 7.65	93.00 + 5.09
8	87.87 + 6.36	78.81 + 8.29	93.81 + 5.43
9	**89.62 + 6.67**	78.62 + 6.98	94.41 + 5.38
10	89.37 + 6.35	77.85 + 7.48	**95.50 + 5.53**

Table 3. Results for energy supplied by the traditional grid.

Units in the hidden layer	PSD	Magnitude peaks	PSD and peaks
1	73.70 + 23.42	47.43 + 7.36	18.30 + 2.79
2	**80.13 + 15.61**	**62.57 + 22.80**	38.58 + 3.50
3	56.77 + 6.54	60.30 + 32.94	51.68 + 6.07
4	58.21 + 5.88	47.43 + 7.36	58.41 + 5.02
5	57.26 + 4.36	50.27 + 8.00	63.25 + 6.80
6	61.75 + 3.22	50.85 + 9.00	67.85 + 4.83
7	62.65 + 5.34	51.21 + 6.30	68.36 + 6.81
8	64.83 + 6.37	50.41 + 6.25	70.23 + 7.14
9	63.15 + 3.77	52.00 + 8.28	71.51 + 6.44
10	66.01 + 3.74	52.50 + 6.98	**73.30 + 7.92**

Table 4. Confusion matrix for the best result when PV system is used.

Class	1	2	3	4	5	6	7	8
1	100	1	0	0	0	0	0	0
2	0	36	0	0	0	0	0	0
3	0	3	40	2	0	0	0	0
4	0	0	0	38	1	0	0	0
5	0	0	0	0	27	0	0	0
6	0	0	0	0	7	34	2	0
7	0	0	0	0	0	3	35	5
8	0	0	0	0	0	0	0	32
No class	0	0	0	0	5	3	3	3

4   Discussion

From information of Fig. 6, it is possible to appreciate that models with five units are enough to discriminate the loads because more units in the hidden layer did not improve the results. These classification rates are upper than 80% for most of the boxes. A similar phenomenon happens for Fig. 7, showing the same effect in the results. In this case, when the traditional grid was used for power generation, the discrimination results were worse than PV system one. In addition, the dispersion of the results was weaker, having the boxes with more scatter for the fifth fold.

For power generation given by PV system, best result was obtained from the PSD information and a model composed by nine units in the hidden layer. Accuracy in this case reached a classification rate of 89%. When information was obtained from the peak value, the accuracy was 94% with a model using two units in the hidden layer. This result evidences that the peak value worked better for discrimination of the studied loads. Finally, when peak and PSD information was used in a simultaneously way, accuracy reached 95%. In this case, the model needed ten units in the hidden layer. This issue can be explained by the employment of 30 features in the ANN input, demanding a bigger architecture to discriminate the loads.

For power generation supplied by the traditional grid, the results were not enough acceptable. The use of PSD allowed to have a discrimination rate of 80% with a model with two units in the hidden layer, but most of the results were between 55% and 65% (see Table 3). Information of the magnitude peak produced an accuracy of 62% as the best, but dropping in a dramatically way after the use of two units in the hidden layer. In addition, when both feature extraction methods were taken into account the accuracy becomes 73%. This comparison indicates that the use of the PV system for electricity generation was better to discriminate the studied loads.

In terms of what load is more difficult to determine, Table 4 provides a direction about it. This table was computed with the best model obtained when the PV system was used and information from PSD and peaks was employed. The table visualizes how the last four classes had worse results in terms of accuracy. Comparing Tables 1 and 4 is trivial to see that loads related with the connection of an induction motor affects the results of discrimination. This can be explained with the increment of noise levels, challenging more the determination task developed by the ANN.

Finally, works with similar methodological proposals have been reported, where accuracies with 90% were reached. In that cases, different acquisition conditions and light modifications in the methods do not allow to compare the present result in a similar way [13]. Also, the state of the art has shown revisions in this area, finding classification rates between 95% and 99% with discrimination models based on ANN, support vector machines and linear discriminant analysis [17]. This final comparison can elucidate how the results presented here were comparable with the findings in the literature, exhibiting a good direction for future approaches.

5 Conclusions

This proposal allowed to establish additional analysis and tools that can provide support in the PQ area, taking as starting point the harmonic distortion and placing a relation of this information with possible nonlinear loads used commonly in homes.

The developed analysis for the PV system employed at SENA is an introduction to begin with complementary studies, where analyses addressed to PQ supervision devices could be of interest.

Specifically talking, the present work found that the use of the PV system in power generation allowed to discriminate in a better way a set of the considered nonlinear loads. Also, the best method to extract features for the load classification was given the use of the magnitude peaks obtained through the FFT for frequency domain analysis.

As future work, it is recommendable to try other techniques for feature extraction and alternative methods for discrimination, and in this way, to establish relation between information obtained from the signal as, for example harmonic distortion and the type of load.

Acknowledgment. Authors want to thank to Universidad Antonio Nariño, that through the Project entitled "Design and implementation of an intelligent system for management of resources in a microgrid supplied by alternative energy" with project code 2017211 and publication code PI/UAN-2018-627GIBIO. In addition, the authors thank to National Learning System (SENA) for supporting this work through the PV system infrastructure used for this study and the metal mechanic center by the collaboration of the instructors involved in this project.

References

1. Hakuta, K., Masaki, H., Nagura, M., Umeyama, N., Nagai, K.: Evaluation of various photovoltaic power generation systems. In: 2015 IEEE International Telecommunications Energy Conference (INTELEC), pp. 1–4 (2015)
2. Messenger, R.A., Abtahi, A.: Photovoltaic Systems Engineering. CRC Press, Boca Raton (2017)
3. Adekol, O.I., Almaktoof, A.M., Raji, A.K.: Design of a smart inverter system for Photovoltaic systems application. In: 2016 International Conference on the Industrial and Commercial Use of Energy (ICUE), pp. 310–317 (2016)
4. Vides-Prado, A., et al.: Techno-economic feasibility analysis of photovoltaic systems in remote areas for indigenous communities in the Colombian Guajira. Renew. Sustain. Energy Rev. **82**, 4245–4255 (2018)
5. Radomes Jr., A.A., Arango, S.: Renewable energy technology diffusion: an analysis of photovoltaic-system support schemes in Medellín. Colombia. J. Clean. Prod. **92**, 152–161 (2015)
6. Hernandez, J.A., Velasco, D., Trujillo, C.L.: Analysis of the effect of the implementation of photovoltaic systems like option of distributed generation in Colombia. Renew. Sustain. Energy Rev. **15**, 2290–2298 (2011)
7. Ortega, M.J., Hernández, J.C., García, O.G.: Measurement and assessment of power quality characteristics for photovoltaic systems: harmonics, flicker, unbalance, and slow voltage variations. Electr. Power Syst. Res. **96**, 23–35 (2013)

8. Benysek, G., Pasko, M.: Power Theories for Improved Power Quality. Springer, Heidelberg (2012). https://doi.org/10.1007/978-1-4471-2786-4
9. Dugan, R.C., McGranaghan, M.F., Beaty, H.W.: Electrical Power Systems Quality. McGraw-Hill, New York (1996)
10. F II, I.: IEEE recommended practices and requirements for harmonic control in electrical power systems, New York, NY, USA (1993)
11. Gil Montoya, F., Manzano-Agugliaro, F., Gómez López, J., Sánchez Alguacil, P.: Power quality research techniques: Advantages and disadvantages. DYNA **79**, 66–74 (2012)
12. Castañeda, A.M.B., Yanchenko, S., Meyer, J., Schegner, P.: Impact of supply voltage distortion on the harmonic emission of electronic household equipment. In: Simposio Internacional sobre la Calidad de la Energía Eléctrica-SICEL (2013)
13. Valtierra-Rodriguez, M., de Jesus Romero-Troncoso, R., Osornio-Rios, R.A., Garcia-Perez, A.: Detection and classification of single and combined power quality disturbances using neural networks. IEEE Trans. Ind. Electron. **61**, 2473–2482 (2014)
14. Raptis, T.E., Vokas, G.A., Langouranis, P.A., Kaminaris, S.D.: Total power quality index for electrical networks using neural networks. Energy Procedia **74**, 1499–1507 (2015)
15. Pedapenki, K.K., Gupta, S.P., Pathak, M.K.: Application of neural networks in power quality. In: 2015 International Conference on Soft Computing Techniques and Implementations (ICSCTI), pp. 116–119 (2015)
16. Saini, M.K., Kapoor, R.: Classification of power quality events–a review. Int. J. Electr. Power Energy Syst. **43**, 11–19 (2012)
17. Khokhar, S., Zin, A.A.B.M., Mokhtar, A.S.B., Pesaran, M.: A comprehensive overview on signal processing and artificial intelligence techniques applications in classification of power quality disturbances. Renew. Sustain. Energy Rev. **51**, 1650–1663 (2015)
18. Barajas, M., Bañuelos-Sánchez, P.: Contaminación armónica producida por cargas no lineales de baja potencia: modelo matemático y casos prácticos. Ing. Investig. y Tecnol. **11** (2010)
19. Pinkus, A.: Approximation theory of the MLP model in neural networks. Acta Numer. **8**, 143–195 (1999)
20. Haykin, S.: Neural Networks and Learning Machines. Prentice Hall, Upper Saddle River (2009)
21. Riedmiller, M., Rprop, I.: Rprop-description and implementation details (1994)
22. Riedmiller, M., Braun, H.: A direct adaptive method for faster backpropagation learning: the RPROP algorithm. In: 1993 IEEE International Conference on Neural Networks, pp. 586–591 (1993)
23. Kubat, M.: Neural Networks: A Comprehensive Foundation by Simon Haykin. Macmillan, New York, 1994 (1999). ISBN 0-02-352781-7

A Systematic Literature Review of Hardware Neural Networks

Dorfell Parra[✉][iD] and Carlos Camargo[iD]

Department of Electric and Electronic Engineering,
Universidad Nacional de Colombia, Bogotá, Colombia
{dlparrap,cicamargoba}@unal.edu.co

Abstract. Although Neural Networks (NN) are extremely useful for the solution of several problems such as object recognition and semantic segmentation, the NN libraries usually target devices which face several drawbacks such as memory bottlenecks and limited efficiency (e.g. GPUs, multi-core processors). Fortunately, the recent implementation of Hardware Neural Networks aims to tackle down this problem and for that reason several researchers had turn back their attention to them. This paper presents the Systematic Literature Review (SLR) of the most relevant HNN works presented in the last few years. The main sources chosen for the SLR were the IEEE Computer Society Digital Library and the SCOPUS indexing system, from which 61 papers were reviewed according to the inclusion and exclusion criteria, and of which after a detail assessment, only 20 papers remained. Finally, the results show that the most popular NN hardware platforms are the FPGAs-based.

Keywords: HNN · SLR · Framework · FPGA · Neural networks

1 Introduction

With the appearance of high performance platforms in the last decade (i.e. GPU, FPGAs), Neural Networks (NNs) are becoming an attractive tool for many classification and object recognition problems. Despite this, the existent APIs for running NNs don't exploit the hardware resources efficiently and for this reason several researches have turn back their attention to Hardware Neural Networks (HNNs) [8,28,30]. HNNs are the implementation of accelerators architectures that evaluate the NNs forward propagation algorithms, by using reconfigurable platforms like FPGAs, or hardcore designs like the VLSI circuits [14,16]. Usually, the networks are trained by means of tools such as MATLAB [2], TensorFlow [4], Microsoft Cognitive Toolkit [3], etc. and then, with the weights and bias calculated, a hardware accelerator is implemented [5,19,31]. Nevertheless, the discussion of how to design HNNs had been widen due to number of works that had been proposed in the last few years [16], and up to day, a new literature

Supported by Universidad Nacional de Colombia.

review is needed to identify the most relevant search streams that had appeared lately, the NN types being used, the organizations leading the research, the research limitations and the quality of the implementation frameworks being proposed. In this paper, we aim to answer all of these questions by means of a Systematic Literature Review (SLR) [10,11]. This paper is organized as follows; Sect. 2 gives a brief background information on HNN and implementation frameworks. Section 3 describes the carried out Systematic Literature Review. Section 4 presents the SLR results and Sect. 5 the discussion. Finally, the conclusion is drawn in Sect. 6.

2   Background

2.1   Hardware Neural Networks (HNNs)

Hardware Neural Networks (HNNs) aim to make the forward evaluation process in a hardware platform [16], by using the weights and biases previously computed with the software libraries in the training process. HNN topologies also receive the name of accelerators, and they are mainly composed by Processing Elements (PEs), memory units and kernel processing units [13,14,17,20,22,23,27], as it is shown in Fig. 1.

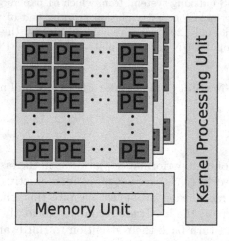

Fig. 1. Neural network accelerator circuit. Adapted from [13,14,17].

Hence, the PEs are hardware cores in charge of computing the neuron basic operations (i.e. floating or fixed point multiplication and addition), the memory units are used to store each neuron output or the activation function when used as a look-up table [17,22,23], and the kernel processing unit manages all the aforementioned resources to compute the NN forward propagation [13].

2.2   HNN Implementation Framework

HNN frameworks intend to integrate the implementation stages (shown in Fig. 2) and reduce the process complexity (e.g. [8, 26, 28]). First, a user interface allows to get the NN model specification (e.g. number of neurons, layers, type of layers, etc.) that will be input in the model abstraction step. Next, with the intermediate representation the framework is going to map the model to the linked libraries (e.g. Torch, TensorFlow, etc.) which will be used to train the NN, create the source code, analyze the NN execution according to predefined metrics and verify and validate the model.

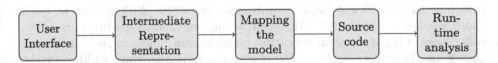

Fig. 2. Basic framework stages. Adapted from [26].

3   Research Methodology

The literature review was made by applying the Systematic Literature Review (SLR) method, proposed by Kitchenham et al. in [10, 11]. The goal of this review is to identify the gaps existing in the HNN implementation process by studying the relevant works in the state of the art. Additionally, according to the literature NN implementations on traditional platforms like general purpose processors don't use hardware resources efficiently, and therefore there is a growing interest in exploring others platforms [28, 30]. On the other hand, Graphical Processing Units (GPUs) are being widely used for training NN because of their high throughput, but those implementations are limited by the local memory available and the communication bandwidth between the host and the GPU [8], [6, 12], besides, its power consumption make them not practicals for embedded systems. Fortunately, these implementations can be improved by using accelerators in custom hardware (e.g. ASICs, FPGA), which have demonstrated to have better performance during the feedforward prediction stage and low power consumption [8]. Moreover, FPGAs are easily programmable and inexpensive, these being why they are the most advantageous option. Furthermore, with the appearance of more optimization design techniques and the amount of FPGA logic resources available being increased, the design exploration space is enlarged boosting the HNN implementation on FPGA-based platforms. For these reasons, this revision of the literature will be focus on HNN implementations on these platforms and will not take into account GPU-based works. Lastly, the SLR method steps are documented below.

3.1   Research Questions (RQ)

The research questions addressed by this review are:

- RQ1: How many frameworks for implementing HNN has been proposed since 2010?
- RQ2: What types of Neural Networks are being addressed?
- RQ3: What individuals and organizations are leading the research?
- RQ4: What are the limitations of current research?
- RQ5: Is the quality of the implementation frameworks improving?

With respect to RQ1, the review starts at 2010 because HNNs start to be a feasible option when the resources and computation platforms available were sufficient for the efficient implementation around 2010. With respect to RQ2, there are several types of NN that can be implemented in hardware, however, differences in input data, classification tasks, neuron function, etc. can lead to choose one instead of another. For this reason it is important to know which are the NN types being used for HNN. With respect to RQ3, it is essential to identify HNN research trends, relevant authors and leading works to be aware of the current problems being studied. With respect to limitations of HNN research (RQ4) the following issues are going to be considered:

- RQ4.1: Were the scope of HNN implementation frameworks limited?
- RQ4.2: Is there evidence that the use of HNNs is limited due to lack of implementation frameworks?
- RQ4.3: Is the quality of implementation frameworks appropriate?
- RQ4.4: Are frameworks contributing to implementation of HNN by defining practice guidelines?

With respect to RQ5, it is important to know if the proposed frameworks are being improved in subsequent works or if the new proposed frameworks are taking different approaches (i.e. where are they going?).

3.2   Search Process

The IEEE Computer Society Digital Library and the SCOPUS indexing system were used in the search process. All searches were based on titles, keywords and abstracts of works published in journals, conferences and symposiums since 2010, which are shown in Table 1. The search string used in the IEEE library was "Neural Networks" AND "Hardware" and the search string used in SCOPUS was TITLE-ABS-KEY("Neural Networks") AND TITLE-ABS-KEY("Hardware") OR TITLE-ABS-KEY("Framework").

3.3   Study Selection

The results for the different searches were added, obtaining a total number of 491 papers published between Jan 1st 2010 and March 31th 2017: 143 from the IEEE digital library, 343 from SCOPUS indexing system and 5 from the open access organizations. To these papers, the following inclusion and exclusion criteria were applied.

Table 1. Selected journals, symposiums and conference proceedings.

Source	Acronym	Organizations	Publication
Transactions on Neural Networks and Learning Systems	TNNLS	IEEE	Journal
Transactions on Very Large Scale Integration Systems	TVLSIS	IEEE	Journal
Transactions on Computers	TC	IEEE	Journal
Neural Networks	NN	ELSEVIER	Journal
Neurocomputing	NC	ELSEVIER	Journal
Information Fusion	IF	ELSEVIER	Journal
Computer Methods and Programs in Biomedicine	CMPB	ELSEVIER	Journal
Pattern Recognition	PR	ELSEVIER	Journal
Engineering Applications of Artificial Intelligence	EAAI	ELSEVIER	Journal
Engineering Research And Development	IJERD	Peer Reviewed	Journal
Electrical and Computer Engineering	IJECE	IAES	Journal
Electronics and Communication Engineering	JECE	IOSR	Journal
International Conference on Data Mining Workshops	ICDMW	IEEE	Conference
International Conference on Computer Science and Network Technology	ICCSNT	IEEE	Conference
International Conference on Architectural Support for Programming Languages and Operating Systems	ASPLOS	IEEE/ACM	Conference
International Conference on Parallel Architectures and Compilation Techniques	PACT	IEEE/ACM	Conference
Annual International Symposium on Computer Architecture	ISCA	IEEE/ACM	Symposium
Annual International Symposium on Field Programmable Custom Computing Machines	ISFPCCM	IEEE	Symposium
Annual International Symposium on Field Programmable Gate Arrays	FPGA	IEEE/ACM	Symposium

Topics used to include the papers:

- Frameworks for implementing HNN with defined research questions, search process, data extraction and data presentation, whether or not the researchers referred to their study as a implementation framework.
- Approach to optimize current implementations of HNN.

Papers on the following topics were excluded:

- Informal literature surveys (no defined research questions; no defined search process; no defined data extraction process).
- Papers presenting implementations of HNN with not defined procedures or not discussing the procedures used.
- Papers presenting GPU-based works, which are limited by local memory constraints and bandwidth communication will also be excluded.
- Duplicate reports of the same study (when several reports of a study exist in different journals the most complete version of the study was included in the review).

After excluding papers that were irrelevant, had not enough information, or were duplicates, there were 56 papers remaining. Those papers were then subject to a more detailed assessment, where each paper was reviewed to identify papers that could be rejected on the basis that they did not include literature reviews, or that they were not related to implementation frameworks. This led to the exclusion of 41 papers. The remaining papers are shown in Table 2.

3.4   Quality Assessment (QA)

Each article was evaluated using the following quality assessment (QA) questions based on [10]:

- QA1: Is the HNN implementation process presented explicitly?
- QA2: Is the literature search likely to have covered all relevant studies?
- QA3: Did the HNN implementation assess the quality/validity of previous included studies?
- QA4: Were the basic data/studies adequately described?

There are three possible outputs for each question with the following score: Y $(yes) = 1$, P $(partly) = 0.5$ and N $(no) = 0$. For QA1: Y the implementation process is presented explicitly, P the implementation process is implicit and N the implementation process is not defined and cannot be readily inferred. For QA2: Y the authors have cited at least 20 works including highly cited works, P the authors have cited between 15 and 19 works including relevant works. N the authors have cited less than 15 works or they have cited irrelevant works. For QA3: Y The HNN implementation performance improved former ones in more than 2x, P the performance was less than 2x of previous ones, and N performance was not reported. For QA4: Y information of each primary study is presented, P each primary study is barely presented, and N any information of each primary study is given.

3.5   Data Collection

The data extracted from each work were: the source (journal, conference or symposium) and full reference, Classification of the study type (i.e. HNN implementation, VLSI design, HNN implementation frameworks), main topic area, the author(s), their institution and the country where it is situated, summary of the study including the main RQ and the answers, research question/issue, quality evaluation, how many primary studies were used in the work.

3.6 Data Analysis

The data was tabulated (see Tables 1 and 3) to show: the number of HNN works published per year and their source (addressing RQ1), whether the HNN work referenced others papers (addressing RQ1), the topics studied by the HNN works, i.e. HNN implementation, VLSI design, HNN implementation frameworks (addressing RQ2 and RQ4.1), the authors: the affiliations of the authors and their institutions was reviewed but not tabulated (addressing RQ3), the number of

Table 2. Systematic review of HNN studies.

Study Ref.	Authors	Date	Paper type	Number primary studies	Review topics
[16]	Misra and Saha	2010	Journal	278	Overview of HNN models: HNN chips, Cellular HNN, Neuromorphic HNN, Optical NN
[9]	Farabet et al.	2011	Conference	27	HNN implementation, hardware architectures
[22]	Abdu-Aljabar	2012	Journal	23	HNN implementation
[27]	Shakoory	2013	Journal	12	HNN implementation
[17]	Mohammed et al.	2013	Journal	9	HNN implementation
[5]	Chen et al.	2014	Conference	44	HNN implementation, VLSI design
[23]	Singh et al.	2015	Journal	11	HNN implementation
[30]	Zhang et al.	2015	Symposium	16	HNN implementation
[31]	Zhou and Jiang	2015	Conference	12	HNN implementation
[7]	Du et al.	2015	Symposium	61	HNN implementation, VLSI design
[28]	Venieris et al.	2016	Symposium	16	HNN implementation framework
[19]	Murakami	2016	Conference	8	HNN implementation
[18]	Motamedi et al.	2016	Conference	10	HNN Parallelism
[8]	Dundar et al.	2016	Journal	49	HNN implementation
[14]	Li et al.	2016	Symposium	11	HNN implementation
[25]	Saldanha et al.	2016	Symposium	13	HNN implementation
[29]	Wang et al.	2016	Symposium	19	HNN implementation, VLSI design
[21]	Ortega-Zamorano et al.	2016	Journal	44	HNN implementation framework for Backpropagation
[13]	Kyrkou et al.	2016	Journal	41	HNN implementation
[15]	Luo et al.	2017	Journal	66	HNN implementation, VLSI design

previous HNN works in each paper (addressing RQ4.2), the quality score for each HNN work (addressing RQ4.3).

4   Results

4.1   Search Results

Including works from journals, symposiums and conferences they were 20 papers reviewed. These papers are shown in Table 2.

4.2   Quality Evaluation of the HNN Works

The HNN works shown in Table 2 were assessed based on the quality assessment (QA) questions. These results are shown in Table 3.

Table 3. Quality evaluation of the HNN studies.

Study Ref.	Paper type	QA1	QA2	QA3	QA4	Total score
[16]	Journal	N	Y	N	Y	2.0
[9]	Conference	Y	Y	Y	P	3.5
[22]	Journal	Y	P	P	Y	3.0
[27]	Journal	Y	N	P	P	2.0
[17]	Journal	N	N	N	N	0.0
[5]	Conference	Y	Y	Y	Y	4.0
[23]	Journal	Y	N	P	N	1.5
[30]	Symposium	Y	P	Y	P	3.0
[31]	Conference	P	P	P	P	2.0
[7]	Symposium	Y	Y	Y	Y	4.0
[28]	Symposium	Y	P	Y	Y	3.5
[19]	Conference	Y	N	P	P	2.0
[18]	Conference	Y	P	Y	Y	3.5
[8]	Journal	Y	Y	Y	Y	4.0
[14]	Symposium	Y	P	P	N	2.0
[25]	Symposium	P	P	P	N	1.5
[29]	Symposium	P	P	P	P	2.0
[21]	Journal	Y	Y	Y	Y	4.0
[13]	Journal	Y	Y	Y	Y	4.0
[15]	Journal	Y	Y	Y	Y	4.0

4.3 Quality Factors

The average Quality Scores (QS) for studies each year, the mean and the standard deviation σ are shown in Table 4. As can be seen the number of HNN studies in the last few years has grew up from 1 study per year up to 9 studies, showing the growing interest for HNN. Also, the average QS per year has been quasi-stable around 3.0 (i.e. 2.88), which can be seen as an increase in the number of most comprehensive works on the topic.

Table 4. Average quality scores for studies by publication date.

	Year							
	2010	2011	2012	2013	2014	2015	2016	2017
Number of studies	1	1	1	2	1	4	9	1
Mean quality score	2.0	3.5	3.0	1.0	4.0	2.625	2.94	4
σ of QS	0.88	0.625	0.12	1.88	1.12	0.255	0.06	1.12

5 Discussion

The answers to the research questions are discussed in this section.

- **RQ1: How many frameworks for implementing HNN has been proposed since 2010?** The revision of several studies from 2010 to 2017 lead to 20 relevant HNN studies. Moreover, it is observed that the interest in HNN is growing up as the number of studies per year. Relevant studies included 1 survey of the HNN implementations proposed before 2010 [16], 4 studies of HNN Very Large Scale Integration (VLSI) designs [5,7,15,29], and the rest of studies proposed an HNN design implemented in a reconfigurable platform (e.g. FPGA). Despite, not all of them presented an explicit framework, the implementation process could be readily inferred.

- **RQ2: What types of Neural Networks are being addressed?** Most of the works aimed to implement Convolutional Neural Networks (CNN) [1, 24]. CNN has been highly accepted in classification and object recognition problems because of its accuracy and relatively fair cost. CNNs are a class of Deep Neural Networks (DNN) where weights are shared across neurons, thus reducing the memory needed to stored the training parameters.

- **RQ3: What individuals and organizations are leading the research?** The leadership of HNNs implementation can be divided by approach. The VLSI design of HNNs is lead by the work group form by the State Key Laboratory of Computer Architecture in China, the Institute of Computing Technology processing in China and the Inria Institution in France. They have designed and implemented at least 4 different HNN chips [5,7,29]. On the other hand, there are different authors that have contributed with several studies about HNNs implementation in reconfigurable platforms. For example, Eugenio Culurciello from the Courant Institute of Mathematical Sciences, New York University and Yann LeCun from the Electrical Engineering Department, Yale University both in USA presented works that include a dataflow processor for vision [9], and an Embedded Streaming DNN Accelerator [8]. In addition, Stylianos Veneris and Christos-Savvas Bougaris from the Department of Electrical and Electronic Engineering, Imperial College in London have presented fpgaConvNet, a framework for mapping CNN on FPGAs in [28]. Finally, Francisco Ortega-Zambrano et al. from the Departamento de Lenguajes y Ciencias de la Computación, Universidad de Málaga in Spain have proposed an efficient implementation of the NN training stage on FPGA in [21]. the mainstreams of HNN.

- **RQ4: What are the limitations of current research?** Currently, due to the number of computational resources (i.e. memory and processing) needed by HNNs implementations, researches aim to commercials FPGA-based systems (i.e. mostly large FPGAs) and the design of HNNs VLSI chips is small. Moreover, there are others factors that restrict the VLSI research such as: the power consumption, die size, fabrication technology and cost. Another important reason is that there is not a general agreement between approaches and there are studies that consider parameters that are not important to others studies, widen the exploration space but reducing the concentrated efforts. Thus, there are a few implementation frameworks proposed and their quality variates from barely acceptable to regular.
- **RQ5: Is the quality of the implementation frameworks improving?** Due to the growing interest in implementing HNN, the number of proposed frameworks by year has been increasing, and so the quality of studies published. For example, current studies offer a more complete description of the proposed work, make comparisons with similar studies and provide external links that widen the information available.

Finally, the following research questions emerge for a future work: Is it possible to implement a NN accelerator to approach different applications?, What are the NN parameters that affect the implementation of NN in hardware platforms?, How is the NN model related to the amount of computational resources needed to implement the HNN in a hardware platform?

6 Conclusions

This chapter presented a Systematic Literature Review of the latest works related to the implementation of neural networks into hardware. Hence, two main streams can be identified: HNNs VLSI designs and HNNs implementation in reconfigurable platforms. Likewise, although the GPU platforms present high throughput in the training stage, they are not suitable for embedded systems or the feedforward stage due to their well known drawbacks (e.g. small local memory, power consumption and communication bandwidth). In addition, the awaken interest in HNNs has revealed the necessity of implementation frameworks, that allow to identify relevant parameters and that present a solid number of stages for accomplish the training of NN in hardware. Unfortunately, frameworks available in the state of the art lack of simplicity, implement the training in software, usually aim to bigger hardware platforms, have an excessive use of logic resources and present an acceptable accuracy. Moreover, there are several problems in the implementation process that still have to be tackle down like: memory bottlenecks, scarce number of resources, complexity of the NNs, implementation precision and accuracy, and efficient HNNs training.

References

1. Face image analysis with convolutional neural networks. https://lmb.informatik. uni-freiburg.de/papers/download/du_diss.pdf. Accessed 15 Feb 2018
2. Mathworks: Matlab. https://www.mathworks.com/products/matlab.html. Accessed 15 Feb 2018
3. Microsoft cognitive toolkit. https://www.microsoft.com/en-us/cognitive-toolkit/. Accessed 15 Feb 2018
4. Tensorflow: An open-source software library for machine intelligence. https://www. tensorflow.org/. Accessed 15 Feb 2018
5. Chen, T., et al.: DianNao: a small-footprint high-throughput accelerator for ubiquitous machine-learning. In: Proceedings of the 19th International Conference on Architectural Support for Programming Languages and Operating Systems - ASPLOS 2014, pp. 269–284, March 2014
6. Chetlur, S., et al.: CuDNN: efficient primitives for deep learning, December 2014. https://arxiv.org/abs/1410.0759. Accessed 15 Feb 2018
7. Du, Z., et al.: ShiDianNao: shifting vision processing closer to the sensor. In: Proceedings of the 42nd Annual International Symposium on Computer Architecture-ISCA 2015, pp. 92–104, June 2015
8. Dundar, A., Jin, J., Martini, B., Culurciello, E.: Embedded streaming deep neural networks accelerator with applications. IEEE Trans. Neural Netw. Learn. Syst. 28(7), 1572–1583 (2017). https://doi.org/10.1109/TNNLS.2016.2545298
9. Farabet, C., Martini, B., Corda, B., Akselrod, P., Culurciello, E., Lecun, Y.: NeuFlow: a runtime reconfigurable dataflow processor for vision. In: IEEE Computer Society Conference on Computer Vision and Pattern Recognition Workshops, pp. 109–116, June 2011
10. Kitchenham, B., Brereton, O.P., Budgen, D., Turner, M., Bailey, J., Linkman, S.: Systematic literature reviews in software engineering - a systematic literature review. Inf. Softw. Technol. 51(1), 7–15 (2008)
11. Kitchenham, B., et al.: Systematic literature reviews in software engineering-a tertiary study. Inf. Softw. Technol. 52, 792–805 (2010)
12. krizhevsky, A.: Survey: implementing dense neural networks in hardware, April 2014. https://arxiv.org/abs/1404.5997. Accessed 15 Feb 2018
13. Kyrkou, C., Bouganis, C.S., Theocharides, T., Polycarpou, M.M.: Embedded hardware-efficient real-time classification with cascade support vector machines. IEEE Trans. Neural Netw. Learn. Syst. 27(1), 99–112 (2016)
14. Li, N., Takaki, S., Tomioka, Y., Kitazawa, H.: A multistage dataflow implementation of a deep convolutional neural network based on FPGA for high-speed object recognition. In: 2016 IEEE Southwest Symposium On Image Analysis and Interpretation (SSIAI), pp. 165–168 (2016). https://doi.org/10.1109/SSIAI.2016.7459201

15. Luo, T., et al.: Dadiannao: a neural network supercomputer. IEEE Trans. Comput. **66**(1), 73–88 (2017)
16. Misra, J., Saha, I.: Artificial neural networks in hardware: a survey of two decades of progress. Neurocomputing **74**(1–3), 239–255 (2010)
17. Mohammed, E.Z., Ali, H.K.: Hardware implementation of artificial neural network using field programmable gate array. Int. J. Comput. Theory Eng. **5**(5), 780–783 (2013)
18. Motamedi, M., Gysel, P., Akella, V., Ghiasi, S.: Design space exploration of FPGA-based deep convolutional neural networks. In: 21st Asia and South Pacific Design Automation Conference, pp. 575–580 (2016). https://doi.org/10.1109/ASPDAC.2016.7428073
19. Murakami, Y.: FPGA implementation of a SIMD-based array processor with torus interconnect. In: 2015 International Conference on Field Programmable Technology, FPT 2015, pp. 244–247, May 2015. https://doi.org/10.1109/FPT.2015.7393159
20. Muthuramalingam, A., Himavathi, S., Srinivasan, E.: Neural network implementation using FPGA issues and application. Inf. Technol. **4**(2), 86–92 (2018)
21. Ortega-Zamorano, F., Jerez, J.M., Munoz, D.U., Luque-Baena, R.M., Franco, L.: Efficient implementation of the backpropagation algorithm in FPGAs and microcontrollers. IEEE Trans. Neural Netw. Learn. Syst. **27**(9), 1840–1850 (2016)
22. Abdu-Aljabar, R.D.: Design and implementation of neural network in FPGA. J. Eng. Dev. **16**(3), 73–90 (2012)
23. Singh, S., Sanjeevi, S., V., S., Talashi, A.: FPGA implementation of a trained neural network. IOSR J. Electron. Commun. Eng. (IOSR-JECE) **10**(3), 45–54 (2015)
24. Saidane, Z.: Image and video text recognition using convolutional neural networks: study of new CNNs architectures for binarization, segmentation and recognition of text images. LAP LAMBERT Academic Publishing (2011)
25. Saldanha, L.B., Bobda, C.: Sparsely connected neural networks in FPGA for handwritten digit recognition. In: Proceedings - International Symposium on Quality Electronic Design (ISQED), pp. 113–117, May 2016
26. Sankaran, A., Aralikatte, R., Mani, S., Khare, S., Panwar, N., Gantayat, N.: DARVIZ: deep abstract representation, visualization, and verification of deep learning models. In: 2017 IEEE/ACM 39th International Conference on Software Engineering: New Ideas and Emerging Results Track (2017)
27. Shakoory, G.H.: FPGA implementation of multilayer perceptron for speech recognition. J. Eng. Dev. **17**(6), 175–185 (2013)
28. Venieris, S.I., Bouganis, C.S.: FpgaConvNet: A framework for mapping convolutional neural networks on FPGAs. In: Proceedings - 24th IEEE International Symposium on Field-Programmable Custom Computing Machines, FCCM 2016, pp. 40–47, May 2016
29. Wang, Y., et al.: Low power convolutional neural networks on a chip. In: IEEE International Symposium on Computer Architecture, no. 1, pp. 129–132, April 2016
30. Zhang, C., Li, P., Sun, G., Guan, Y., Xiao, B., Cong, J.: Optimizing FPGA-based accelerator design for deep convolutional neural networks. In: Proceedings of the 2015 ACM/SIGDA International Symposium on Field-Programmable Gate Arrays - FPGA 2015, pp. 161–170, February 2015
31. Zhou, Y., Jiang, J.: An FPGA-based accelerator implementation for deep convolutional neural networks. In: 4th International Conference on Computer Science and Network Technology (ICCSNT), pp. 829–832, December 2015

Computational Intelligence

On Computing the Variance of a Fuzzy Number

Juan Carlos Figueroa-García(✉) ⓘ, Miguel Alberto Melgarejo-Rey, and José Jairo Soriano-Mendez

Universidad Distrital Francisco José de Caldas, Bogotá, Colombia
{jcfigueroag,mmelgarejo,josoriano}@udistrital.edu.co

Abstract. This paper presents a comparison of three well known methods for computing the variance of a fuzzy number and a proposal based on the Yager index for convex fuzzy sets: the Carlsson-Fullér, Mendel-Wu, and the sample variance of a fuzzy set. Some considerations about the obtained results are provided and some recommendations are given.

Keywords: Fuzzy numbers · α-cuts · Fuzzy variance · Yager index

1 Introduction and Motivation

Fuzzy sets are able to deal with non-probabilistic uncertainty when statistical data is absent/incomplete which is commonly seen in several engineering and mathematical problems. Both the mean and variance are important order statistics of a population (as seen in the probabilistic approach), so for a fuzzy granule (understood as an entity characterized by a linguistic label and a membership function) its centroid is usually taken as its expected value and its variance is seen as a spread measure regarding its centroid.

This way, we focus on analyzing the proposal of Figueroa-García, Melgarejo-Rey and Soriano-Méndez [8] who used the Yager index (see Yager [16]) for convex fuzzy sets to compute its variance. Other interesting works focused on the centroid/mean and variance of a fuzzy set were proposed by Carlsson and Fullér [1], Klir and Folger [10], Klir and Yuan [11], Mendel and Wu [12], Wu and Mendel [15], Figueroa-García and Pachon-Neira [6,7], and Figueroa-García, Chalco-Cano and Román-Flores [5], so we compare all of their results to see their differences.

The paper is divided into six sections. Section 1 introduces the topic. In Sect. 2, some basics on fuzzy sets/numbers are provided; in Sect. 3, some definitions of the centroid/mean of a fuzzy set are presented. Section 4 presents the concept of variance of a fuzzy set from different authors perspectives; Sect. 5 presents some application examples, and Sect. 6 presents the concluding remarks of the study.

2 Basics of Fuzzy Sets/Numbers

Let $\mathcal{P}(X)$ be the class of all crisp sets, $\mathcal{F}(X)$ is the class of all fuzzy sets, $\mathcal{F}_1(X)$ is the class of all convex fuzzy sets, and $I = [0,1]$ be the set of values in the

© Springer Nature Switzerland AG 2018
A. D. Orjuela-Cañón et al. (Eds.): ColCACI 2018, CCIS 833, pp. 89–98, 2018.
https://doi.org/10.1007/978-3-030-03023-0_8

unit interval. A fuzzy set namely A is characterized by a membership function $\mu_A : X \to I$ defined over a universe of discourse $x \in X$. Thus, a fuzzy set A is the set of ordered pairs $x \in X$ and its membership degree, $\mu_A(x)$, i.e.,

$$A = \{(x, \mu_A(x)) \mid x \in X\}. \tag{1}$$

Let us denote $\mathcal{F}_1(\mathbb{R})$ as the class of all fuzzy numbers. Henceforth we refer to A as a fuzzy number, as shown as follows:

Definition 1. *Let $A : \mathbb{R} \to I$ be a fuzzy subset of the reals. Then $A \in \mathcal{F}_1(\mathbb{R})$ is a Fuzzy Number (FN) iff there exists a closed interval $[x_l, x_r] \neq \emptyset$ with a membership function $\mu_A(x)$ such that:*

$$\mu_A(x) = \begin{cases} c(x) & \text{for } x \in [c_l, c_r], \\ l(x) & \text{for } x \in [-\infty, x_l], \\ r(x) & \text{for } x \in [x_r, \infty], \end{cases} \tag{2}$$

where $c(x) = 1$ for $x \in [c_l, c_r]$, $l : (-\infty, x_l) \to I$ is monotonic non-decreasing, continuous from the right, i.e. $l(x) = 0$ for $x < x_l$; $l : (x_r, \infty) \to I$ is monotonic non-increasing, continuous from the left, i.e. $r(x) = 0$ for $x > x_r$.

The α-*cut* of a set $A \in \mathcal{F}_1(\mathbb{R})$, $^\alpha A$ is the set of values with a membership degree equal or greatest than α, this is:

$$^\alpha A = \{x \mid \mu_A(x) \geqslant \alpha\} \; \forall \; x \in X, \tag{3}$$

$$^\alpha A = \left[\inf_x {}^\alpha \mu_A(x), \sup_x {}^\alpha \mu_A(x) \right] = [\check{A}_\alpha, \hat{A}_\alpha]. \tag{4}$$

An open question around the variance of a fuzzy set is: Is it a distance? Can we computed it using classical statistics? To do so, we firstly provide some definitions of the mean/centroid of a fuzzy set to then refer to the variance of a fuzzy number (see Figueroa-García et al. [8]) and several classical approaches, as shown in next sections.

3   Yager Index, Carlsson-Fuller Mean and Centroid of a Fuzzy Number

3.1   Yager Index of a Fuzzy Number

Yager [16] has proposed one of the most important and widely applied ranking methods for convex fuzzy sets. As Chaudhuri and Rosenfeld [2] and Hung and Yang [9] defined regarding α levels for computing distances, the Yager Index for convex fuzzy sets comes from the idea of integration of the arithmetic mean for every $^\alpha A$, namely $M(A_\alpha)$:

$$M(A_\alpha) = \frac{\check{A}_\alpha + \hat{A}_\alpha}{2} \; \forall \, \alpha \in [0, 1], \tag{5}$$

$$I(A) = \int_{[0,1]} M(A_\alpha) \, d\alpha = \frac{1}{2} \int_{[0,1]} (\check{A}_\alpha + \hat{A}_\alpha). \tag{6}$$

so the index $I(A)$ for discrete α is

$$I(A) = \frac{1}{2} \sum_{i=1}^{n} (\check{A}_{\alpha_i} + \hat{A}_{\alpha_i}) \Delta_{\alpha_i}, \tag{7}$$

where $\{\check{A}_\alpha, \hat{A}_\alpha\}$, $\alpha \in [0,1]$ is the α-level of A, and Δ_{α_i} is a delta step on α.
If $X = \text{supp}(A)$, and $x \in X$ is discrete where $x_i < x_{i+1}$, then we have:

$$M(X) = \frac{1}{n} \sum_{i=1}^{n} x_i. \tag{8}$$

Definition 2. *Let us define the mass of a fuzzy set A, $E(A)$ as follows:*

$$E(A) = \int_x x \cdot \mu_A(x)\, dx.$$

So if A is symmetric, then $I(A)$ can be easily re wrote as:

$$\frac{1}{|A|} E(A) = I(A),$$

and the centroid of any symmetric fuzzy number equals its Yager index i.e.

$$I(A) = \frac{1}{|A|} E(A) = C(A).$$

Roughly speaking, $I(A)$ is a Lebesgue integral (a Riemann integral in the discrete case) done over α instead of $x \in X$, since its domain is smaller and easier to measure in the case of fuzzy numbers. $I(A)$ has some desirable properties such as reflexivity, symmetry, and transitivity which are important properties that allows to use it as a reliable measure of A.

3.2 Carlsson-Fullér Possibilistic Mean of a Fuzzy Number

One of the most popular approach has been proposed by Carlsson and Fullér [1] who originally proposed a method for computing the possibilistic mean of a fuzzy set. Thus, the possibilistic mean of A, $m(A)$ is as follows:

$$m(A) = \int_{[0,1]} \alpha(\check{A}_\alpha + \hat{A}_\alpha)\, d\alpha. \tag{9}$$

3.3 Centroid of a Fuzzy Number

The well known centroid of a fuzzy A, $C(A)$ is as follows:

$$C(A) = \frac{\int_x x \cdot \mu_A(x)\, dx}{\int_x \mu_A(x)\, dx} = \frac{1}{|A|} \int_x x \cdot \mu_A(x)\, dx, \tag{10}$$

and its discrete version is:

$$C(A) = \frac{\sum\limits_{i=1}^{n} x_i \cdot \mu_A(x_i)}{\sum\limits_{i=1}^{n} \mu_A(x_i)}. \tag{11}$$

All three $I(A), m(A), C(A)$ can be used as expectations of A, so the computation of the variance of A depends on what measure is selected as expectation, as shown as follows.

4   Variance of a Fuzzy Number

There are several approaches to measure the variance of a fuzzy set/number. In this paper we refer to two important approaches to measure the variance of a fuzzy number: the Carlsson-Fullér approach (see Carlsson and Fullér [1]) and the Mendel-Wu approach (Mendel and Wu [12] and Wu and Mendel [15]) and the proposal of Figueroa-García et al. [8].

4.1   Carlsson-Fullér Possibilistic Variance of a Fuzzy Number

Carlsson and Fullér [1] proposed a method for computing the possibilistic variance of a fuzzy set. Thus, the possibilistic variance of A, $Var(A)$ is as follows:

$$Var(A) = \frac{1}{2} \int_{[0,1]} \alpha (\hat{A}_\alpha - \check{A}_\alpha)^2 \, d\alpha. \tag{12}$$

This definition is based on measuring the range of $^\alpha A$ as a Minkowski L_2 distance and it works only for convex fuzzy sets.

4.2   Mendel-Wu Variance of a Fuzzy Number

Mendel and Wu [12] proposed another method for computing the variance of a fuzzy set, based on its centroid $C(A)$. Then, the variance of A, $v(A)$ is as follows:

$$v(A) = \frac{1}{n} \sum_{i=1}^{n} [x_i - C(A)]^2 \mu_A(x_i). \tag{13}$$

Finally, the α-cut version of $v(A)$ is as follows:

$$v(A) = \frac{1}{2n} \sum_{i=1}^{n} \left[(\check{A}_\alpha - C(A))^2 + (\hat{A}_\alpha - C(A))^2 \right]. \tag{14}$$

where $\{\check{A}_\alpha, \hat{A}_\alpha\}$, $\alpha \in [0,1]$ is the α-level of $A \in \mathcal{F}_1(\mathbb{R})$.

4.3   Yager-Based Variance of a Fuzzy Number

Now, the variance of A based on the Yager index (see Figueroa-García et al. [8]) is defined as follows.

Definition 3. *Let $A \in \mathcal{F}_1(\mathbb{R})$ be a fuzzy number, then the **variance** of A regarding $I(A)$ namely $V(A)$ is as follows:*

$$M_2(A_\alpha) = \frac{1}{2}\left[(I(A) - \check{A}_\alpha)^2 + (\hat{A}_\alpha - I(A))^2\right], \tag{15}$$

$$V(A) = \int_x M_2(A_\alpha)\, d\alpha, \tag{16}$$

which results into the following α-representation:

$$V(A) = \frac{1}{2}\int_{[0,1]}(I(A) - \check{A}_\alpha)^2 d\alpha + \frac{1}{2}\int_{[0,1]}(\hat{A}_\alpha - I(A))^2\, d\alpha \tag{17}$$

so the variance for the discrete case is

$$V(A) = \frac{1}{2}\sum_{i=1}^{n}\left[(I(A) - \check{A}_{\alpha_i})^2 + (\hat{A}_{\alpha_i} - I(A))^2\right]\Delta_{\alpha_i} \tag{18}$$

where $\{\check{A}_\alpha, \hat{A}_\alpha\}$, $\alpha \in [0,1]$ is the α-level of A and Δ_{α_i} is a delta step on α.

5   Application Examples

In order to compare all three methods, we selected triangular, trapezoidal and Gaussian fuzzy sets for which we computed 1000 equidistant α-cuts which returned $\check{A}_\alpha$ and $\hat{A}_\alpha$. Additionally, we computed the sample mean $\bar{x}$ and variance S^2 for $\check{A}_\alpha$ and $\hat{A}_\alpha$ considered as realizations of A, as shown as follows:

$$\bar{x} = \frac{1}{2n-1}\sum_{i=1}^{n}\left[\check{A}_\alpha + \hat{A}_\alpha\right], \tag{19}$$

and the sample variance is:

$$S^2 = \frac{1}{2n-1}\sum_{i=1}^{n}\left[(\bar{x} - \check{A}_\alpha)^2 + (\hat{A}_\alpha - \bar{x})^2\right], \tag{20}$$

where $\{\check{A}_\alpha, \hat{A}_\alpha\}$, $\alpha \in [0,1]$ is the α-level of $A \in \mathcal{F}_1(\mathbb{R})$.

The obtained results for all methods are summarized into Tables 1, 2, and 3.

5.1   Triangular Fuzzy Number

Figure 1 shows a triangular fuzzy number $T(\check{a}, \bar{a}, \hat{a})$.

Table 1 shows 10 different choices of the parameters of a triangular fuzzy number. Some differences between them are easy to see: while the proposed Yager mean/variance is equal to the sample mean/variance, the Carlsson-Fullér and Mendel-Wu approaches are different.

Also note that $I(A) = m(A) = C(A) = \bar{x}$ for symmetric triangular fuzzy numbers i.e. $|\bar{a} - \check{a}| = |\hat{a} - \bar{a}|$.

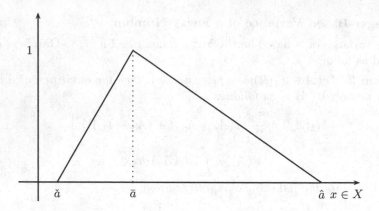

Fig. 1. Triangular fuzzy set/number $\tilde{A}$

Table 1. Obtained results for triangular fuzzy numbers

	Parameters			Yager		Carlsson-Fullér		Mendel-Wu		Sample	
#	$\check{a}$	$\bar{a}$	$\hat{a}$	$I(A)$	$V(A)$	$m(A)$	$Var(A)$	$C(A)$	$v(A)$	$\bar{x}$	S^2
1	0	5	10	5	8.33	5	4.17	5	2.08	5	8.34
2	0	5	20	7.5	35.42	6.67	16.67	6.67	8.85	7.5	35.46
3	0	5	30	10	83.33	8.33	37.50	8.33	20.83	10	83.43
4	0	5	50	15	241.67	11.67	104.17	11.66	60.42	15	241.96
5	10	30	50	30	133.33	30	66.67	30	33.33	30	133.47
6	10	35	50	32.5	135.42	33.33	66.67	33.34	33.85	32.5	135.56
7	10	40	50	35	141.67	36.67	66.67	36.67	35.42	35	141.82
8	10	45	50	37.5	152.08	40.00	66.67	40.01	38.02	37.5	152.26
9	5	20	35	20	75.00	20	37.50	20	18.75	20	75.08
10	10	20	30	20	33.33	20	16.67	20	8.33	20	33.37

5.2 Trapezoidal Fuzzy Number

Figure 2 shows a trapezoidal fuzzy number $T(\check{a}, \bar{a}_1, \bar{a}_2, \hat{a})$.

Table 2 shows 10 different choices of the parameters of a trapezoidal fuzzy number. It is interesting to see that the proposed Yager mean/variance is equal to the sample mean/variance (as in triangular fuzzy numbers), and the Carlsson-Fullér and Mendel-Wu approaches shows some differences.

Again, note that $I(A) = m(A) = C(A) = \bar{x}$ for symmetric trapezoidal fuzzy numbers i.e. $|\bar{a}_1 - \check{a}| = |\hat{a} - \bar{a}_2|$.

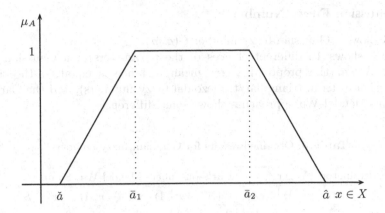

Fig. 2. Trapezoidal fuzzy set/number $\tilde{A}$

Table 2. Obtained results for trapezoidal fuzzy numbers

	Parameters				Yager		Carlsson-Fullér		Mendel-Wu		Sample	
#	$\check{a}$	$\bar{a}_1$	$\bar{a}_2$	$\hat{a}$	$I(A)$	$V(A)$	$m(A)$	$Var(A)$	$C(A)$	$v(A)$	$\bar{x}$	S^2
1	0	5	20	25	12.5	102.08	12.5	85.42	12.5	42.69	12.5	102.14
2	0	5	20	50	18.75	302.60	16.67	194.79	16.66	101.67	18.75	302.83
3	0	5	20	100	31.25	1094.27	24.99	569.79	24.99	323.74	31.25	1095.35
4	0	5	45	50	25	508.33	25	470.83	25	235.38	25	508.59
5	0	30	45	50	31.25	302.60	33.33	194.79	33.34	101.67	31.25	302.83
6	0	40	45	50	33.75	256.77	36.67	128.12	36.67	72.52	33.75	257.03
7	35	40	45	50	42.49	27.50	42.5	18.75	42.5	9.37	42.48	27.93
8	5	15	20	30	17.50	64.61	17.5	39.58	17.5	19.78	17.50	64.69
9	0	15	20	35	17.5	118.75	17.5	68.75	17.5	34.34	17.5	118.85
10	0	25	30	35	22.5	127.08	24.17	68.75	24.17	37.12	22.5	127.20

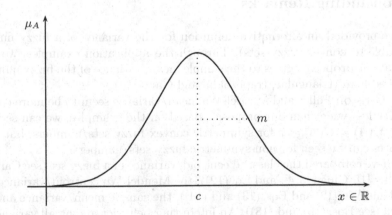

Fig. 3. Gaussian fuzzy number A

5.3 Gaussian Fuzzy Number

Figure 3 shows a Gaussian fuzzy number $G(c, m)$.

Table 3 shows 10 different choices of the parameters of a Gaussian fuzzy number. Again, the proposed Yager mean/variance is equal to the sample mean/variance (as in triangular/trapezoidal fuzzy numbers), and the Carlsson-Fullér and Mendel-Wu approaches shows some differences.

Table 3. Obtained results for Gaussian fuzzy numbers

	Parameters		Yager		Carlsson-Fullér		Mendel-Wu		Sample	
#	c	m	$I(A)$	$V(A)$	$m(A)$	$Var(A)$	$C(A)$	$v(A)$	$\bar{x}$	S^2
1	0	1	0	2.01	0	1	0	0.5	0	2.02
2	0	2	0	8.02	0	4	0	2	0	8.07
3	0	3	0	18.05	0	9	0	4.5	0	18.16
4	0	4	0	32.08	0	16	0	8	0	32.29
5	0	5	0	50.13	0	25	0	12.5	0	50.45
6	0	6	0	72.18	0	36	0	18	0	72.64
7	0	7	0	98.25	0	49	0	24.5	0	98.88
8	0	8	0	128.32	0	64	0	32	0	129.14
9	0	9	0	162.41	0	81	0	40.5	0	163.45
10	0	10	0	200.51	0	100	0	50	0	201.79

Again, note that $I(A) = m(A) = C(A) = \bar{x}$ for Gaussian fuzzy sets since they are symmetric.

6 Concluding Remarks

We have provided an alternative definition for the variance of a fuzzy number (applicable to convex fuzzy sets). Through the application examples, we have seen that our proposal agrees to the sample mean/variance of the fuzzy numbers considered here (triangular, trapezoidal and Gaussian).

The Carlsson-Fullér and Mendel-Wu mean/variance seem to be more conservative i.e. less wide than our proposal. Based on the examples, we can see that $I(A) = M(A) = C(A) = \bar{x}$ for symmetric convex fuzzy sets/numbers, but some differences can be seen for non-symmetric fuzzy sets/numbers.

We have compared the classical centroid/variance of a fuzzy set (see Carlsson and Fullér [1], Chiao [3,4], and Eq. (12)), the Mendel-Wu centroid/variance (see Mendel and Wu [12] and Eqs. (13) and (14)), the sample mean/variance and our proposal (see Eqs. (16) and (18)). An interesting behavior among all variances is that the Mendel-Wu approach is a half the value of the Carlsson-Fullér approach which is turn a half of the Yager/sample variance for symmetric fuzzy numbers.

Finally, the proposed variance seems to converge to the sample mean/variance which meets with Kolmogorov theory. Note that sample mean/variance does not weight any value x_i and they are shown to be unbiased/consistent/efficient estimator for the first and second moments of any random variable, so the proposed Yager variance has the potential to meet those desirable properties.

6.1 Further Topics

Further theoretical analyses and generalization are required in order to provide a better understanding of our proposal. Its extension to Interval Type-2 fuzzy sets are promissory fields to be covered (see Figueroa-García, Chalco-Cano and Román-Flores [5], and Salazar-Morales et al. [13, 14]).

References

1. Carlsson, C., Fullér, R.: On possibilistic mean value and variance of fuzzy numbers. Fuzzy Sets Syst. **122**(1), 315–326 (2001)
2. Chaudhuri, B., RosenFeld, A.: A modified hausdorff distance between fuzzy sets. Inf. Sci. **118**, 159–171 (1999)
3. Chiao, K.P.: A new ranking approach for general interval type-2 fuzzy sets using extended alpha cuts representation. In: International Conference on Intelligent Systems and Knowledge Engineering, vol. 1, pp. 594–597. IEEE (2015)
4. Chiao, K.P.: Ranking type-2 fuzzy sets using parametric graded mean integration representation. In: International Conference on Machine Learning and Cybernetics, vol. 1, pp. 606–611. IEEE (2016)
5. Figueroa-García, J.C., Chalco-Cano, Y., Román-Flores, H.: Yager index and ranking for interval type-2 fuzzy numbers. IEEE Trans. Fuzzy Syst. (1), 1–9 (2017, in Press)
6. Figueroa-García, J.C., Pachon-Neira, D.: On ordering words using the centroid and Yager index of an interval Type-2 fuzzy number. In: Proceedings of the 2015 Workshop on Engineering Applications (WEA), vol. 1, pp. 1–6. IEEE (2015)
7. Figueroa-García, J.C., Pachon-Neira, D.: A comparison between the centroid and the Yager index rank for type reduction of an interval Type-2 fuzzy number. Rev. Ing. Univ. Dist. **2**, 225–234 (2016)
8. Figueroa-García, J.C., Soriano-Mendez, J.J., Melgarejo-Rey, M.A.: On the variance of a fuzzy number based on the Yager index. In: Proceedings of IEEE ColCaCi 2018, pp. 1–6. IEEE (2018)
9. Hung, W.L., Yang, M.S.: Similarity measures between type-2 fuzzy sets. Int. J. Uncertain. Fuzziness Knowl.-Based Syst. **12**(6), 827–841 (2004)
10. Klir, G.J., Folger, T.A.: Fuzzy Sets, Uncertainty and Information. Prentice Hall, Upper Saddle River (1992)
11. Klir, G.J., Yuan, B.: Fuzzy Sets and Fuzzy Logic: Theory and Applications. Prentice Hall, Upper Saddle River (1995)
12. Mendel, J.M., Wu, D.: Cardinality, fuzziness, variance and skewness of interval type-2 fuzzy sets. In: Proceedings of FOCI 2007, pp. 375–382. IEEE (2007)
13. Salazar-Morales, O., Serrano-Devia, J.H., Soriano-Mendez, J.J.: Centroid of an interval type-2 fuzzy set: continuous vs. discrete. Revista Ingeniería **16**(2), 1–9 (2011)

14. Salazar-Morales, O., Serrano-Devia, J.H., Soriano-Mendez, J.J.: A short note on the centroid of an interval type-2 fuzzy set. In: Proceedings of 2012 Workshop on Engineering Applications (WEA), pp. 1–6. IEEE (2012)
15. Wu, D., Mendel, J.M.: A comparative study of ranking methods, similarity measures and uncertainty measures for interval type-2 fuzzy sets. Inf. Sci. **179**(1), 1169–1192 (2009)
16. Yager, R.: A procedure for ordering fuzzy subsets of the unit interval. Inf. Sci. **24**(1), 143–161 (1981)

eHealth Services Based on Monte Carlo Algorithms to Anticipate and Lessen the Progress of Type-2 Diabetes

Huber Nieto-Chaupis[✉]

Center of Research eHealth, Universidad de Ciencias y Humanidades,
Av. Universitaria 5175, Los Olivos, Lima39, Peru
huber.nieto@gmail.com

Abstract. We present a computer-based eHealth system expected to provide tele-consults aimed to reduce complications due to the diabetes disease in adult population mainly between 30 and 60 years old. The software of the tele-consultations system which is essentially based in probabilities is entirely based in the Monte Carlo technology. This stochastic method is supported with a mathematical model which is build through acquired data that allows us to model and carry out predictions on the glucose's values in time within a certain statistical error. The idea behind of this eHealth system is the rapid identification of those people with a potential risk to acquire complications derived from the high values of glucose in time. The conclusion derived from this study supports the fact that opportune intervention derived from the tele-consultations might alleviate and to improve the diabetes treatment by employing simple low-cost mobile phones and minimal software applications. We illustrated the prospective implementation of this tele-care system with simulations for people with an old diagnosis of diabetes and demonstrating the prospective role o these eHealth systems aimed to improve the quality of life in the middle and long term. From a combined sample composed by acquired data and Monte Carlo, 3 from 4 diabetes patients might be keeping a desirable control of their glucose's values with a continuous assistance of an eHealth system.

Keywords: Monte Carlo · Teleconsult · Type-2 diabetes

1 Introduction

IIt's well known that the diabetes disease in their forms type-I and type-II [1] constitutes a worldwide interest for the public health operators based on the fact that there is progressive increasing of the number of diagnosis by which most of them might be to have started a progressive deterioration of certain physiological functionalities since diabetes is a systemic disease affecting diverse functionalities of the human physiology [2].

© Springer Nature Switzerland AG 2018
A. D. Orjuela-Cañón et al. (Eds.): ColCACI 2018, CCIS 833, pp. 99–110, 2018.
https://doi.org/10.1007/978-3-030-03023-0_9

Commonly when people has a diabetes diagnosis the main task is that of facing a drastic change in the lifestyle fact which can be done progressively in time. Clearly this demands a discipline in foods and an adequate pharmacology [3] in a proper manner.

In large cities where public hospitals [4] might no be capable to attend people massively, exists the possibility to implement tele consultations by which people is required to be trained in order to optimize the relation doctor-patient.

A different case is the one of the first attention level where might be located far from hospitals or large health centers. Therefore the so-called teleconsultation might operate also in these centers and subsequently to derive patients to hospital or similar health centers [5]. Thus, the operativeness of a tele consultation as a call of voice of video would depend essentially in their main software and hardware components.

Although a standard hardware is crucial, software based in open source algorithms might be of great importance since it might be optimized and improved continuously. Indeed, these softwares might be used also to carry out predictions in time consistently.

In this paper we focus in the computational design of a tele consult system entirely based on Monte Carlo algorithms. The system also would have capabilities to make coherent predictions about the glucose values in order to detect the potential cases which might to acquire complications [6].

Basically we focus on the behavior of glucose in time from a data taken in the past and with this we build a mathematical model inserted in the Monte Carlo algorithms. Thus we have paid attention to the cases where people is surpassing the established ranges prescribed by the doctor.

Normally are these cases where the teleconsultation [7,8] would demonstrate its effectiveness as to provide a clear vision to the health specialists in order that give them position to make a solid intervention [9].

We have focused on the most simple scheme of data acquisition by which the people would send through text message their glucose's values in order to be stored in a data server that process the received information of patient and simultaneously identifying those cases that require urgent intervention [10].

With the accumulated data the software implemented in the sever makes calculations about the possible values of glucose [11] particularly in those critic cases where a doctor needs to arrange a interview in the very short term either in the first attention level or hospital.

Once the critic cases have been identified, the health personal carry out a continuous tracking to them in the sense of guaranteeing the surveillance of the identified patient until previous to their interview with doctor.

Clearly the teleconsultation is not restricted to doctor but also to psychologists for instance and other health professional that can add more schemes of recovery as to improve the health of patient to aim to surpass critic phases during the disease progress.

In second section we present all the quantitative machinery: the mathematical model based from a sample of real data of patients in Lima city. From this sample we propose a curve that would describe the evolution of glucose in time. In third section we use the Monte Carlo technique to simulate additional samples aimed to test the functionality of the proposal eHealth system[14]. In fourth section we present the results of the Monte Carlo analysis. Finally regarding the simulation results of the paper the conclusion is drawn.

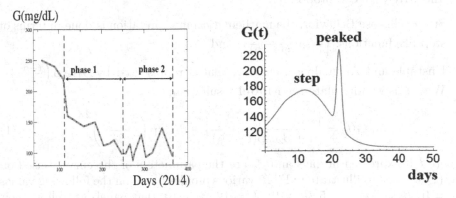

Fig. 1. Left: resulting average curve from a sample of 10 adults patients taken during the 2014. Right: a possible curve of glucose behavior in a type-2 diabetes patient showing two phases: (i) the step-like phase and (ii) a the peaked distribution.

2   Formulation of Model

2.1   Description of Samples and Data

Our study has as starting point the data from 10 patients whose ages are ranging between 30 and 60 y.o. whose locations are along peripheral areas de Lima city and by which most of them have manifested jumps on their glucose values due to multiple factors mainly in the lifestyle of patient. On the other hand, all of them are pursuing a well-defined pharmacology based on metformin and similar strategies. Doses of metformin is ranging between 250 mg and 1000 mg. In some cases approximately 4 of them are following a therapy consisting in metformin together to vildagliptin as well. Data was acquired from fasting glucose test achieved by a glucometer. Glucose test was done in an inhomogeneous way, each 5 days or strictly weekly. Left panel of Fig. 1 displays the resulting average for this sample of 10 patients randomly selected from a more large population of patients. Thus in a first instance we can see that the curve is clearly composed by two phases well-defined. Phase-1: According to Fig. 1 left panel the first phase is fully understood in terms of the pure effect of antidiabetics on the glucose of patients. Phase-2: Behavior beyond 250th day is showing peaks and jumps fact that is translated as the lack of a firm strategy of either patient or public health operators to maintain an optimum value of glucose despite of the fact of the intake of antidiabetic is achieved.

2.2   Mathematical Approach

A fitting on the data (left panel Fig. 1) yields approximately a composition of up two different morphologies, one which is perceived as a step-like function and one which is a composition of various Gaussian profiles since the portions of the data along phase 2 is displaying various peaks.

Intuitively we can propose

- Stable Glucose Behavior: the mathematical interpretation is done in terms of step-like functions: $\left[\frac{\beta_1}{1+\beta_2\exp(t-\beta_3)}\right]$, and

- Unstable and Anomalous Behavior, that is represented by $\beta_4\mathrm{Exp}\left(\frac{t-\beta_5}{\beta_6}\right)^2$. With this we formulate the **Model** resulting in

$$G(t) = \left[\frac{\beta_1}{1 + \beta_2\exp(t - \beta_3)}\right] + \beta_4\mathrm{Exp}\left(\frac{t - \beta_5}{\beta_6}\right)^2 \tag{1}$$

where t is expressed in days and β_j are the parameters of the curve. Based on Eq.(1) we pass to illustrate in Fig. 2 various profiles based on the following values $\beta_1 = 10, \beta_2 = 0.1, \beta_3 = 5, \beta_4 = 10, \beta_5 = 12, \beta_6 = 10$. (left panel) as well as their respective variations of β on the right panel. Here we have introduced an strong oscillating function denoting the nonlinearity of the glucose behavior in time.

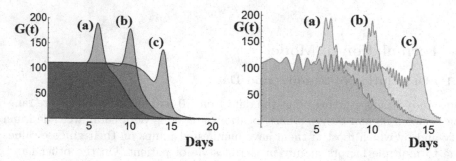

Fig. 2. Left: different manifestations of the Eq. (1) with their parameters varying inside the ranges defined by $\Delta\beta_j$. Right: same as top panel but with the insertion of a non-linear function of the β parameters.

3   Monte Carlo Algorithms: Generation and Prediction

3.1   Data Generation

Once the pattern is identified as seen in Fig. 1, the subsequent step is the data simulation through the Monte Carlo method. Clearly the testing of the teleconsultation requires to include more samples that allows us to make prediction within the allowed statistical errors.

Thus, the central objective of this step in the generation of $G_q(x)$ curves, with q the number of patients. It demands to provide random fluctuations on the β_j parameters as seen in right side of Fig. 2. We have assigned a maximum window of 20% for all β_j, in order to avoid any anomalous Monte Carlo generation in the resulting $G_q(x)$ curves. Special attention is paid on the β_2 parameter since it enhances the nonlinearity of curve. One important condition to reject the $G_\ell(x)$ for a ℓ-individual during the process of Monte Carlo generation, is the absence of a two-phases behavior.

A brief description of the algorithm used to produce extra samples with the Monte Carlo method is given below: Firstly we adjudicate initial values to the parameters β_j and take the initial value of the bin. We start the loop over 100 samples. Therefore for each point x_n as done in line-8 we evaluate Eq. 1 (line-9) that include the inclusion of the values of β_j as well. In line-10 we provide values for the random numbers $c_{n,j}^q$ and $\theta_{n,j}^q$ which are selected from a RNG (Random Number Generator). In line-12 the acceptance or rejection is carried out through the comparison of a couple of random numbers. When the acceptance is done then we pass to add an increment on the updated values of the parameters (line-13). When the Monte Carlo steps have reached their maximum value the we turn

---

Algorithm 1. Monte Carlo Samples Generation

---

1 **INITIALIZATION**
2 ASSIGN VALUES β_j
3 INITIAL BIN x_0
4 **FOR** $q \subset [0,100]$ Calculates G_q
5 ASSIGN TO q: GENDER, AGE, WEIGHT
6   **FOR** $n \to 1$ **TO** $\mathcal{N}$-bin **DO**
7   **FOR** $j \to 1$ to 6 **DO**
8   $x_n = x_n + x_{n-1}$
9   $G_q[x_n, \beta_{n,j}^q]$
10   $c_{n,j}^q$ and $\theta_{n,j}^q$
11   **FOR** $\ell \to$ (MC-STEPS) **DO**
12   if $c_{n,j}^q > \theta_{n,j}^q$ **THEN**
13     $\beta_{n,j}^q \to \beta_{n,j}^q + \delta\beta_{n,j}^q$
14   **IF** $\ell \to \ell_{\text{MAX}}$ **THEN**
15   **MC AVERAGES**
16   **MC ERRORS**
17      $G_q[x_n, \beta_{n,j}^q] \to G_q[x_n, \beta_{n,j}^q + \delta\beta_{n,j}^q]$
18   if $|G_q[x_n, \beta_{n,j}^q + \delta\beta_{n,j}^q]| > \bar{G}_{\text{TH}}$ **THEN**
19   Rejects $G_q[x_n, \beta_{n,j}^q + \delta\beta_{n,j}^q]$
20        $\ell \to \ell + 1$
21   **END IF ANDEND FOR**
22 **UNTIL** $q=100$

---

to estimate the errors attached to the gathered statistical. With these errors we pass to estimate the value of glucose for the time x_n together to the updated values $\delta\beta_{n,j}^q$. In order to keep a close tracking between the Monte Carlo outputs and the theoretical value as given in Eq. (1) any strong deviation between Monte Carlo and theoretical model is rejected (line-18) since it might not be reflecting the pattern as basis for the data generation. Rejection demands to opt next loop as seen in the line-20. For this exercise has demanded to employ 1000 Monte Carlo steps. therefore 100 extra samples were generated. These generated data are used to test the e-health system.

3.2   Glucose Concentration Prediction

Once data have been generated, the next step to be taken is the prediction and interpretation of glucose concentration behavior for subsequent days. The central idea here is to be able to compute the prediction for the glucose values deviations during the next N-days. In this manner, the maximum and minimum glucose values allows us to adjudicate priorities to urgent access to teleconsults over the next weeks. Of course this list of priorities is updated depending upon the incoming glucose tests arriving to the server. In order to evaluate the effect of the teleconsultation it's mandatory to estimate the fail or success of the disease treatment with the following relation:

$$P_F^i = \frac{G_{q,\text{MAX}}^i - G_{q,\text{MIN}}^i}{G_{q,\text{MEAS}} - G_{q,\text{IDE}}} << 1. \tag{2}$$

Thus, for a glucose concentration behavior prediction in a time T_h, the following proposition is presented and demonstrated,

Proposition 1. Consider the probability of success and fail P_S and P_F of the diabetes treatment in time where $P_S + P_F = 1$.

For a q-patient, we consider $G_q(t_i)$ as glucose history's continuous curve in time represented by Eq. 1 where $t_{\text{MEA}} < t_i < \infty$ and t_{MEA} time of glucose measurement occurrence, then the fail probability $< P_F >$ of treatment of diabetes for a q patient is given by

$$< P_F >_q^{k+1} = \frac{1}{N} \sum_{i=1}^{N} \frac{\eta^{i,k} G_{q,\text{MAX}}^{i,k} - \gamma^{i,k} G_{q,\text{MIN}}^{i,k}}{G_{q,\text{MEAS}} - \rho^{i,k} G_{q,\text{IDE}}} \tag{3}$$

where the nomenclature $<>$ means the average. η^i and γ^i factors to enhance $G_{q,\text{MAX}}^{i,k}$ and $G_{q,\text{MIN}}^{i,k}$ maximum and minimum glucose values respectively ρ^i enhances $G_{q,\text{IDE}}$ the ideal glucose value extracted from Eq. 1. Roughly speaking for all times of glucose measurement $< P_F >_q^{k+1} + < P_S >_q^{k+1} = 1$, for a discrete time $k + 1$.

Proof. Consider glucose value $G_{q,\text{MEAS}}$ for a time where no any prescription strategy is applied.

Then the reference quantity is defined as the difference between $G_{q,\text{MEAS}}$ and $G_{q,\text{IDE}}$. Consider now two events already measured $G_{q,\text{MAX}}$ and $G_{q,\text{MIN}}$ when a treatment is already running. Thus we can obtain the differences given by $G_{q,\text{MAX}} - G_{q,\text{MIN}}$ and $G_{q,\text{MEA}} - G_{q,\text{ID}}$.

As result the quotient between these two differences is defined as the full fail of treatment. However, the case of importance is when $G_{q,\text{MAX}} - G_{q,\text{MIN}} \ll G_{q,\text{MEAS}} - G_{q,\text{IDE}}$ whose quotient is defined as the fail probability for the q-patient with a running treatment.

$$P_F^i = \frac{\eta G_{q,\text{MAX}}^i - \gamma G_{q,\text{MIN}}^i}{G_{q,\text{MEA}}^i - \rho G_{q,\text{IDE}}^i}. \tag{4}$$

Thus we can associate the upper index k to those values which have been extracted by the server in a random manner to measure an instantaneous value t_k. This demonstrates the Proposition 1.

For example, consider the case when a patient have accumulated data in the server. So we take in a random manner the values of $G_{\text{MAX}} = 250$ and $G_{\text{MIN}} = 210$, and for a set values of β, γ and ρ then $G_{\text{MEA}} = 300$ and $G_{\text{IDE}} = 250$, then $P_F^i = \frac{250-210}{300-250} = 0.80$ or 80% which is certainly a high probability to fail the treatment. All values are expressed in mg/dL.

Therefore for a time t_i the success of the treatment combining interviews and teleconsultations can be written as

$$P_S^i = 1 - \frac{G_{q,\text{MAX}}^i - G_{q,\text{MIN}}^i}{G_{q,\text{MEA}}^i - G_{q,\text{IDE}}^i}. \tag{5}$$

Once the formalism is defined then we are able to estimate the future probabilities P_S^{i+1}, P_S^{i+2},..., P_S^{i+s-1}, and P_S^{i+s}. Clearly one should be noted that the success of the teleconsultation must be proportional to the decreasing of the glucose values in those people that have started their eHealth-based consults with values above the allowed.

3.3 Priorities in the eHealth Scheme

The immediate application of previous proposition is the construction of a list of priorities or hierarchy of levels. In fact, the resulting quantity that the proposition yields the possible predicted values of glucose: $\Lambda^T = [P_{1,F}^T, P_{2,F}^T, ..., P_{q_1-1,F}^T, P_{q_1,F}^T]$ and for a subsequent time $T+1$ the new vector is written as $\Lambda^{T+1} = [P_{1,F}^{T+1}, P_{2,F}^{T+1}, ..., P_{q_2-1,F}^{T+1}, P_{q_2,F}^{T+1}]$. It is important to remark that $P_{q_1,F}^T \neq P_{q_2,F}^{T+1}$ since q_2 and q_1 denote different patients. In addition, for a time $T+T_G$, we have $\Lambda^{T+T_G} = [P_{1,F}^{T+T_G}, P_{2,F}^{T+T_G}, ..., P_{q_G-1,F}^{T+T_G}, P_{q_G,F}^{T+T_G}]$ where $P_{1,F}^{T+T_G} > P_{2,F}^{T+T_G} > .... > P_{q_G-1,F}^{T+T_G} > P_{q_G,F}^{T+T_G}$. Thus the quality of the future values (in the sense of putting down those high values) might be determined by the rapid medical intervention which is arranged with the updated predictions of glucose values. The Algorithm 2 describes the scheme of teleconsults that are driven by the Eq. (1) and proposition-1 essentially. It is assumed that it is a code written in Java, C++ o another language programming which requires input values. As is seen the algorithms firstly pick ups all receives values assumed to be sent by the patients for a time which can be days or until weeks. When data have been accumulated then the algorithm makes an ordering of the values along the pat. So that a vector containing the information from the pas values is constructed (line-5). For a new arrival we define the present (line-6) as just the latest value that was successfully registered by the system. The reader can note that we are using as template the Eq. (1) and their parameters obtained in a random manner. The code working together to Algorithm 1 is taken those values as the ones of line-2 of it. Therefore we estimate the future values $\mathcal{G}_r$ by knowing the latest value $\mathcal{G}_0$. By reaching the maximum value of loop in r, the vector of the future values in next days is explicitly given as seen in line-11. Once it is estimated the crucial part of this algorithm is that of identifying the maximum values of glucose: i.e., to identify whom will have the highest glucose values in order to arrange a teleconsultation. According to the predicted values as seen in line-15 is done the vector of priorities we can assign a teleconsult line-19 with its quality of service Γ_l. In line-20 we estimate again the vector of glucose values per patient which is assumed is different since it has to have the effect of the teleconsultation in conjunction to the rapid intervention

4   Simulation Results

The results are shown in Fig. 3. Of 100 samples plus 10 real data we have selected the ones which have been selected with the marked difference of the effect of the teleconsults resulting to be in a range between 25 and 40 patients. For example in the top panel is perceived the effect of the teleconsults where the pink color denotes the ones which have had teleconsults and rapid intervention in according to line-20 of Algorithm 2. While the oranges ones denote the cases where not any intervention is done and only pick up the data as estimated in line-15. Here we can perceive that there are patients that were decreasing their glucose values in around two weeks with the support of the teleconsultation and anticipated intervention. In the middle panel we can see the accumulation of events showing low values of glucose when the eHealth system is working over 20 days continuously. The fact that the patients have opted only by teleconsult without receiving intervention appears to be a latent risk for them as seen in the apparition of events (orange cubes) with values up 100 mg/dL over 20 days of surveillance. For these events the fail probability has turned out to be high reaching a 85% in average, fact indicates the limitations of the eHealth system. Here is important to note the relevance of the eHealth system as a tool that persuade the patient to contact a doctor in the short term. In the bottom panel is shown the case where solely are counted the cases that received teleconsults and intervention for values of glucose between 170 mg/dL and 200 mg/dL. Actually the bottom histogram also reveal us that the ordering that makes the Algorithm 2 is subject to random fluctuations from the fact that the Monte Carlo drives the

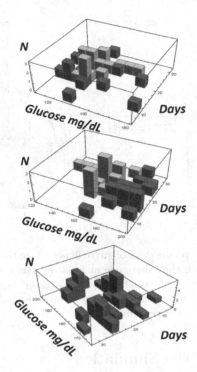

Fig. 3. Monte Carlo simulation of the number of patients (z-axes) versus glucose values and days of surveillance (x-y plane). Pink and orange cubes denote the events where it was done teleconsults and intervention and only teleconsults respectively. (Color figure online)

processing of arrived data to the server. We assume a minimal infrastructure as to the cell phones of users so they can send continuously the calls and being received in a unstoppable manner by the server. Under the assumption that the wireless call operator provides a dedicate network to the eHealth appears the question if the networking architecture is the fully responsible to maintain an acceptable quality of service. In Fig. 4 we can see the possible effect of the eHealth service to detain the progress of diabetes. In left panel up to 25 patients have achieved to reach values less than 150 mg/dL whereas in right panel 8 of the total sample were identified with low values with less than 125 mg/dL. By assuming that a 50% have successfully reached a value of 100 mg/dL and the respective Monte Carlo error then 3 of 4 might to surpass possible episodes of jumps of glucose values. These claims are based in the curve that is adjusted well to the data.

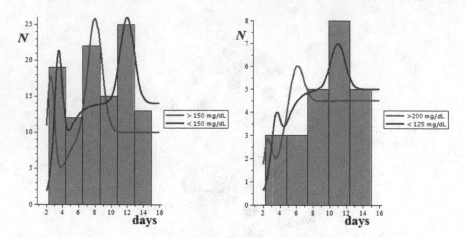

Fig. 4. Number of patients versus days of surveillance. Top: the case where a fit adjusts the data (blue line) yielding that there are at least 36 patients with glucose less than 150 mg/dL for 15 days. Bottom panel denotes the same but we identified up to 8 patients with their glucose below 125 mg/dL for two weeks of surveillance as seen in the fitting (blue line) (Color figure online)

5   Consistency of the Simulations

In Fig. 5 are shown up to 4 different scenarios of different values for the parameters β, γ and ρ basically ranging between 0.2–0.8, 0.1–0.9, and 0.3–0.5 respectively. All histograms have been filled under the assumption that the Eq. 1 dictates the possibly "true" glucose behavior. Thus we have plotted the number of patients versus the expected difference "MEA-IDE" in units of mg/dL. All these plots correspond to the cases where the teleconsultations have running during two weeks. In top left panel is seen the case where the distribution of patients is homogeneous for the differences ranging between 0 and 10 mg/dL. The mean value of patients have been calculated to be of order of 10 individuals. From this plot for instance for "MEA-IDE" between 0 and 5 we can estimate that at least 50 patiens have kept their values of glucose near to the ideal or theoretical ones. This values doubles in the top right plot. In the bottom plots are shown the probably the cases where most of the patients have not reached objectives as suggested by the endocrinologist and we note that at least 15 (or up to 20) patients have showing jumps in their values of glucose as seen in the difference MEA-IDE that turns out to be of order of up to 100 mg/dL. It means for instance whether IDE = 100 then MEA ≈ 200 mg/L fact that indicates a possible fail of the services of eHealth or also en in the side of the patients: the abandon of a healthy lifestyle the assures the stability of the glucose values on the ones which are permanently suggested by the eHealth specialists. The error bars seen in all plots of Fig. 5 have exhibited an error of order of 10%.

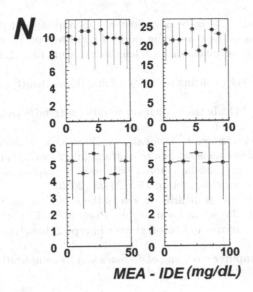

Fig. 5. Monte Carlo simulation of the number of patients exhibiting the difference between the measured "MEA" and ideal "IDE" values expressed in mg/dL, with up to for 4 scenarios of values of the parameters γ, β, and ρ.

6   Conclusion

In this study we have used Monte Carlo algorithms as part of a eHealth service aimed to analyze and anticipate the progress of diabetes. Our simulations combine real and simulated data and the results have been obtained with an error of order of 10%. This can be translated as the fact that 3 of 4 patients would maintain their standard glucose's values. Clearly it is also perceived as the necessity of providing and effective attention to those patients which are not following a strict control of their glucose values for large terms. It demands to apply other alternative methodologies which would have to be adjustable to the social, culture and educational aspects for the human group under study.

References

1. Vistisien, D., et al.: Patterns of obesity development before the diagnosis of type 2 diabetes: the Whitehall II cohort study. Plos Med. **11**(2), e1001602 (2014)
2. Hanefel, M.: Use of insulin in type 2 diabetes: what we learned from recent clinical trials on the benefits of early insulin initiation. Diabetes Metab. **40**, 391–399 (2014)
3. Megías, M.C.: Influence of macrolides, nutritional support and respiratory therapies in diabetes and normal glucose tolerance in cystic brosis. A retrospective analysis of a cohort of adult and younger patients. Diabetes Metab. Syndr. Clin. Res. Rev. **9**, 1–6 (2015)
4. Chen, L.: Effect of lifestyle intervention in patients with type 2 diabetes: a meta-analysis. Metab. Clin. Exp. **64**, 338–347 (2015)

 5. Sullivan, T., et al.: The co-management of tuberculosis and diabetes: challenges and opportunities in the developing world. PLoS Med. **9**(7), e1001269 (2012)
 6. Kamsu, B.: Systemic modeling in telemedicine. Eur. Res. Telemed. **3**(2), 57–65 (2014)
 7. Kropf, N., et al.: Telemedicine for older adults. Home Health Care Serv. Q. **17**(4), 1–11 (1999)
 8. Kim, T.H., Lee, H.H.: Is telemedicine a worldwide trend? Arch. Gynecol. Obstet. **289**(5), 925–926 (2014)
 9. Meneghini, L., Kesavadev, J., Demissie, M., Nazeri, A., Hollander, P.: Oncedaily initiation of basal insulin as add-on to metformin: a 26-week, randomized, treat-to-target trial comparing insulin detemir with insulin glargine in patients with type 2 diabetes. Diabetes Obes. Metab. **15**, 729–36 (2013)
10. Jeon, C., et al.: Diabetes mellitus increases the risk of active tuberculosis: a systematic review of 13 observational studies. PLoS Med. **5**(7), e152 (2008)
11. Rispin, C.M.: Management of blood glucose in type 2 diabetes mellitus. Am. Fam. Phys. 79(1) (2009)
12. Palmer, A.J.: Computer modeling of diabetes and its complications. Diabetes Care **30**(6), 1638–1646 (2007)

Predicting Student Drop-Out Rates Using Data Mining Techniques: A Case Study

Boris Pérez[1,2(✉)] ⓘ, Camilo Castellanos[2] ⓘ, and Darío Correal[2] ⓘ

[1] Univ. Francisco de Paula Stder., Cúcuta, Colombia
borisperezg@ufps.edu.co
[2] Universidad de los Andes, Bogotá, Colombia
{cc.castellanos87,dcorreal}@uniandes.edu.co

Abstract. The prevention of students dropping out is considered very important in many educational institutions. In this paper we describe the results of an educational data analytics case study focused on detection of dropout of Systems Engineering (SE) undergraduate students after 6 years of enrollment in a Colombian university. Original data is extended and enriched using a feature engineering process. Our experimental results showed that simple algorithms achieve reliable levels of accuracy to identify predictors of dropout. Decision Trees, Logistic Regression, Naive Bayes and Random Forest results were compared in order to propose the best option. Also, Watson Analytics is evaluated to establish the usability of the service for a non expert user. Main results are presented in order to decrease the dropout rate by identifying potential causes. In addition, we present some findings related to data quality to improve the students data collection process.

Keywords: Student drop out · Student desertion prediction
Educational data mining · Prediction models

1 Introduction

In every country, education is synonymous with rapid economic growth [13]. Universities can play a key role as builders of qualified human capital and innovation systems in their countries [5]. Also, they help supporting the economic growth of a country [13]. High level of education along with more students completing their studies are the required conditions to improve level of human capital in the society. The completion of university is correlated to the increase of life expectancy, increase in social status, reduction of the risk of unemployment, among others [10].

There has been increasing interest from universities in understanding the behavior of successful students. The reputation of these institutions is measured by the percentage of students who graduate and by the strategies the university has to retain its students [16]. The early identification of students who are at risk of dropping out is critical for the success of any retention strategy [16]. Then

© Springer Nature Switzerland AG 2018
A. D. Orjuela-Cañón et al. (Eds.): ColCACI 2018, CCIS 833, pp. 111–125, 2018.
https://doi.org/10.1007/978-3-030-03023-0_10

it becomes necessary to detect these students as early as possible, maintaining intensive and continuous intervention to reduce the dropout levels [15,18].

The identification of dropout rates has become a major issue for Universities. In Colombia, there is an initiative from the Ministerio de Educación called SPADIES (System for Prevention and Analysis of Desertion in Institutions of Higher Education) [11]. This initiative was designed by the Center for Economic Studies (SEDE) at the University of the Andes to follow up on the problem of dropout in higher education, to calculate the risk of desertion of each student, and to classify these students by groups. This initiative can support the evaluation of strategies for each of the situations that influence dropout such as student status, academic program and institution; and also promotes the consultation, consolidation, interpretation and use of this information (tables and graphs, each by various criteria).

In the literature, researches have been investigating the finding of the characteristics of both the student and his context influencing his decision to drop out of university. Tinto's model [21] is the most widely accepted model in student retention literature. Tinto concluded that the student dropout is strongly related to their degree of academic (grade performance and intellectual development) and social integration (peer-group interactions and faculty interactions) at university. Bharadwaj and Pal [3,4] identified that attributes of the score in a senior secondary exam, residence, various habits, annual family income, and family status were shown to be important parameters for dropout. Additionally, Kovacic [14] identified that enrollment data could be used to predict dropout. Variables like students attendance in class, hours spent studying after class, family income, mother's age and mother's education are significantly related to student dropout. Specifically, it has been found that the factors like mother's education and family income are highly correlated with student performance [9].

However, there is still no consensus in the literature on the causes of dropout in universities. Despite the existence of multiple studies on university dropout, there is a little research related to the computer science area. Also, the vast majority of studies are focused on static variables, leaving aside the dynamic component of the grades the student obtains during his studies. It is fundamental to investigate the causes that lead students of Systems Engineering program to withdraw the course before its completion.

Thus, considering the factors that influence the dropout rates in universities, the aim of the present study is to answer the following questions: (a) what are the key determinants of undergraduate student dropout rates in a Systems Engineering program of a Colombian Private University? And, (b) what data mining technique is more suitable to find these key determinants? To do so, we model student dropout using data gathered from academic databases from 2004 to 2010. In addition, this paper presents the differences between using a programmatic approach to identify the key determinants versus the automatic approach offered by Watson Analytics from IBM. We do not take into account data related to the enrollment process like demographic information. Our approach considered a significant population of a private university.

This paper is organized as follows: Sect. 2 reviews the related work. Section 3 introduces the data mining methodology. Section 4 describes the data set used. Section 5 presents the data preparation. Section 6 describes the modeling process. Section 7 reports results. Section 8 offers a preliminary proposal to use these results in a real context. Finally, Sect. 9 outlines the conclusions.

2 Related Work

Data mining is the area which analyzes huge repositories of data to extract important patterns, association and relations among all these and is therefore a valuable tool for converting data into usable information [9]. Data mining can discover hidden information in various domains, including marketing, banking, educational research, surveillance, telecommunications fraud detection, and scientific discovery. Education is one of these domains where the primary concern is the evaluation and, in turn, enhancement of educational organizations [20].

Educational Data Mining (EDM) is a discipline engaged with the develop of methods and techniques not only for exploring and analyzing the data that come from educational context but also for extracting hidden information for better understanding students. This information can be used in several educational processes such as predicting course enrollment, estimating student dropout rate, detecting atypical values in students' transcripts, and improvement of student models that would predict student's characteristics or academic performances [20, 23].

Tinto's model [21] is the most widely accepted model in student retention literature. Tinto concluded that the decision of students to persist or drop out of their studies is strongly related to their degree of academic and social integration at university. In the academic system, Tinto analyzed the grade performance and intellectual development. In the social system, peer-group interactions and faculty interactions were also analyzed. However, Brunsden et al. in [6] tested the model with path analysis using LISREL8 software and concluded that Tinto's model may not be the most appropriate for dropout research. They

Bharadwaj and Pal [4] used EDM to evaluate student performance among 300 students from five different colleges who were enrolled in an undergraduate computer course. They employed a Bayesian classification scheme of 17 attributes, of which the score in a senior secondary exam, residence, various habits, annual family income, and family status were shown to be important parameters for academic performance. In a second study, Bharadwaj and Pal [3] constructed a new data set which included student attendance, and test, seminar, and assignment grades in order to predict academic performance. A similar study was proposed by Kovacic [14], who applied EDM to identify which enrollment data could be used to predict student academic performance. In this study, he used CHAID and CART algorithms on a dataset of student enrollment.

In another study, Al-Radaideh et al. [1] analyzed student's academic data (student gender, student age, student department, high school grade, lecturer degree, lecturer gender, among others) building a classification model using the decision tree method to improve the quality of the higher educational system. They found that high school grade was the attribute with the highest gain ratio and was considered the root node of the decision tree. The Holdout method and the K-Cross-Validation method (k-CV) were used to evaluate the model. However, they found that the collected samples and attributes were not sufficient to generate a classification model of high quality.

Gerben et al. [8] conducted a case study in which they used machine learning techniques to predict student success using features extracted from student pre-university academic records. Their experimental results showed that simple and intuitive classifiers (i.e.: decision trees) gave useful results with accuracies between 75% and 80%. One of their findings was that the strongest predictor of success was the grade for the Linear Algebra course, which was not seen as the decisive course.

Despite these studies, it's not clear which data mining algorithms are preferable in this context. For example, Luan in [12] built predictors using clustering as means of data exploration and classification. In [17], Romero and Ventura presented a survey on EDM where one of their findings was that association analysis has become a popular approach. Finally, Herzog in [19] presented the results of a case study where Bayesian networks and neural networks were outperformed by decision tree algorithms but on small educational datasets.

3 Methodology

The analytic task performed in this work is a binary classification task where dropout (0, 1) is the target or dependent variable. To do this, we follow the Cross Industry Standard Process for Data Mining methodology (CRISP–DM) [7]. This methodology is useful for planning, communicating the project team, and documenting. It provides a generic check-lists which advises the steps to be taken and provides practical advice for all steps. The life cycle of a data mining project [22] is broken down into six phases as presented in Fig. 1.

The phases are described below. *Business understanding* allows definition of the business goal, in our case the student dropout phenomenon, covered in the previous sections. *Data understanding* involves data collection, identification of data quality problems and discovering of insights. *Data preparation* covers feature extraction, data wrangling, and can require multiple iterations. *Modeling* consists of technique selection, application and calibrating parameters. There is a close link between Data Preparation and Modeling. Often, one realizes data problems while modeling and gets ideas for constructing new data. *Evaluation* is focused on the performance assessment of the models built in the previous phase. *Deployment* deals with the operationalization of the model within the real context. These phases will be tackled in the following sections.

For data understanding, data preparation and modeling, we use an open source tool in Python as our data science development environment: Jupyter

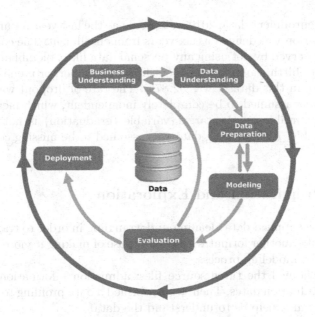

Fig. 1. Cross industry standard process for data mining methodology

notebook[1], Python data analysis library (Pandas[2]) to deal with data structures, Scikit-learn[3] (machine learning library), Seaborn[4] a statistical data visualization tool and Graphviz[5] a graph visualization software to generate the decision tree charts.

4   Dataset

The data set used in this study comes from 762 students enrolled in the Systems Engineering Program at a private university in Bogotá, Colombia. The data is organized in three tables, and together they include the following features:

- Admission information, including minimum demographic information (gender, birth date, marital status).
- Graduation dates, including date of graduation and the academic program.
- Transcript records including the courses taken and the grades for each of them, the academic program and the academic cumulative average.

Our focus was on students entering the university from first term of 2004 through the second term of 2010. Although the institutional databases has the

---

[1] http://jupyter.org.
[2] http://pandas.pydata.org.
[3] http://scikit-learn.org.
[4] http://seaborn.pydata.org.
[5] http://graphviz.org.

latest student enrollment data, 2010 was chosen as the last year for analyses since student graduation was defined as six years from enrollment. The confidentiality of data was preserved by not using any personal data like Colombian national ID number, date of birth, campus wide identification number, or name. The overall graduation rate in this dataset was 52.87%. The rate of dropout was 47.13%.

The data was assumed to be completely independent, which means that the effect of each variable on the target variable (graduation) is not affected by any other variable. Also, missing data was assumed to be missing completely at random.

5   Data Preparation and Exploration

In this phase, we applied data cleaning and wrangling in order to transform original data set into another format with the purpose of making it more appropriate and valuable for modeling process.

Firstly, we joined the three source files: admission information, transcript records and graduation dates. Then, we performed a data profiling to get descriptive statistics which help us to understand the data.

The approach we used is grouping courses within similar academic field (systems engineering (SE), mathematics (MATH), physics (PH), management (MGMT), Language (LAN), Biology (BIO), etc.). For this, we aggregated the course grades, and course repetitions by student and by faculty. Additionally, we added the student age at the enrollment time and the standard deviation of academic term cumulative average to reflect the variance (irregularity) of academic performance. As a result, we obtained an aggregate dataset with 31 columns which contain the following relevant fields: gender, marital status, age at enrollment, academic terms (by academic field), academic cumulative average, standard deviation of academic term averages, course grades averages, course repetitions by faculty, and the dropout indicator (0,1).

Figure 2 presents the grade averages by academic fields, segmenting by dropout indicator. As expected, dropout students present lower grade averages across all subject. Management classes are the most challenging for both types of students, followed by mathematics and physics.

Categorical fields such as sex and marital status were transformed to numeric values because the most of models require numeric inputs. To do this transformation, we used dummy variables to represent each categorical value as a binary indicator column. Also, we scaled out the features to normalize the magnitudes and prevent that high magnitude fields skew the feature's weights into the machine learning models.

Although sampled from a diverse university, Fig. 3 shows a clear dominant demographic profile for SE students: single males between 17 and 19 years old, accounting for 71.5% of the dataset.

The correlation between significant variables can be visualized in Fig. 4. As expected, we find strong positive correlation between the averages of the different subjects, with mathematics and physics being the highest. We also have

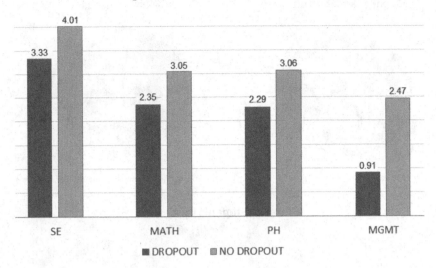

Fig. 2. Average grades by subject group of all students at the end of their program.

that GPA is most influenced by system engineering classes (since compose the majority of classes in the program) and has a negative correlation with the target variable. Surprisingly, there is a strong correlation between the amount of times a student fails a system engineering and a biology class.

As mentioned before, the frequency of dropouts in the dataset is almost equal (52.87% for *no* and 47.13% for *yes*) so no over/under sampling is required.

Finally, since it is of interest for school administrations to detect dropout as early as possible, we include the average grade progression for the first three semesters for both types of students in Fig. 5. Selected subjects where Mathematics, Physics and System Engineering since the represent the majority of courses in the program.

6 Modeling

According to previous works described in Sect. 2, models offering the best accuracy [2,8,14] are: Decision Tree, Logistic Regression and Naive Bayes. In addition, we used Random Forest model as a complemented technique because it is suitable for classification and regression, and it operate by constructing a multitude of decision trees at training time and outputting the class that is the mean prediction of the individual trees. All models where applied to data restricted to the first, second, third and last semester after enrollment.

We applied Cross Validation (CV) in every modeling tasks to avoid overfitting and we used all dataset in training and testing steps. In the approach called k-*fold* CV, the training set was split into k smaller sets, in our case, k=5 folds. And for each fold:

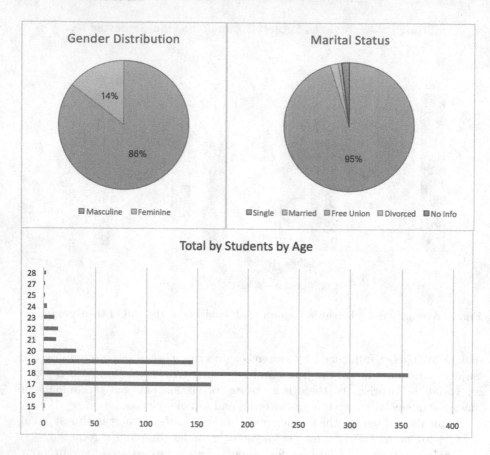

Fig. 3. Student's sex, marital status and age summary.

- A model is trained using *k–1* (4) of the folds as training data (4/5 of data).
- The resulting model is validated with the resting part of the data (1/5).

6.1   Models Setup

- **Decision Tree.** We trained a decision tree model with *gini* criteria and CV mentioned before. The model without pruning contains twelve levels and fifty-one leaf nodes. This tree model was pruned using max depth parameter = 4 levels and we got 7 leaf nodes tree, as shown in Fig. 6.
- **Logistic Regression.** We trained the logistic model with the following parameters: tolerance for stopping criteria = 0.0001, inverse of regularization strength = 1.0, and solver = *liblinear*.
- **Naive Bayes.** We used a Gaussian Naive Bayes algorithm.
- **Random Forest.** We trained this model with the following parameters: number of trees in the forest = 10, maximum depth of the tree = 4 and random state = 0, and a cross-validation generator = 6.

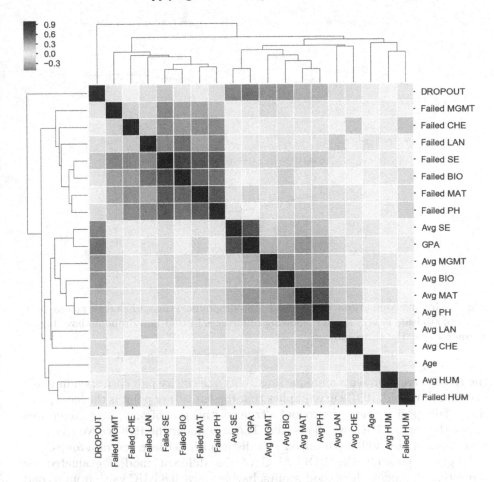

Fig. 4. Cluster map of feature's correlations

6.2 Watson Analytics

Watson Analytics is a smart service for analyzing and visualizing data to quickly discover patterns and meaning in data, without having any previous knowledge. Watson Analytics use guided data discovery, automated predictive analytics and cognitive capabilities to interact with data to get findings you understand. No previous configuration is required.

7 Model Evaluation and Results

In this section, models selected were evaluated in terms of Receiver Operating Characteristic (ROC) curve analysis. In a ROC curve the true positive rate, also called Sensitivity, is plotted in function of the false positive rate (i.e.: 100-Specificity) for different cut-off points of a parameter. Each one of the points on

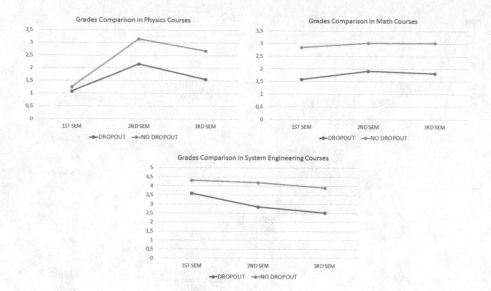

Fig. 5. Grading comparison: dropouts vs. non-dropouts. Tangent behavior among both types of students is very similar, showing differences in that non-dropouts have higher average across the three subjects.

the ROC curve represents a unique sensitivity/specificity pair corresponding to a decision threshold. ROC curve implies that top left corner point is the ideal point (i.e. a false positive rate of zero and a true positive rate of one), so, a larger area under the curve (AUC) will be better. The area under the ROC curve (AUC) is a measure of how well a parameter can distinguish between several groups.

Figure 7 presents the ROC–AUC of the different models evaluated by semester. All models show good results, having above 0.8 AUC value from second semester onwards and as expected, they all show that the longer the student is in enrolled the better the prediction gets. Random Forest shows the best results, giving an 0.91 AUC on the third and a 0.97 AUC on the last semester, making it the ideal model for this case study.

For the Decision Tree model, we found that:

- Case 1: a student with *Avg SE* lower or equal than 3.505, and *Failed SE* lower than or equal to 0.1389 (scaled 0 to 5), and *GPA* lower than or equal to 3.5164, has 100% (143/143) for dropping out.
- Case 2: a student with *Avg SE* lower than or equal to 3.505 and *Failed SE* lower than or equal to 0.1389 (scaled 0 to 5) and *GPA* greater than 3.5164, has 88.8% (48/54) for dropping out.

For the Logistic Regression model, the significant coefficients are displayed in Table 1. As expected, we have that *Failed MAT* and *Failed MGMT* increase the probability of dropping out, opposed to *Avg SE* that reduces the risk of not completing the degree. Surprisingly, *Failed SE* has a negative influence on the

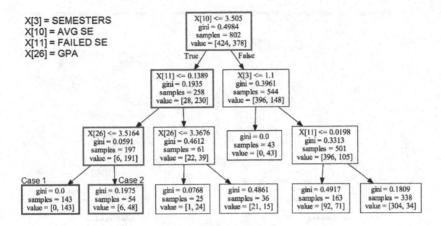

X[3] = SEMESTERS
X[10] = AVG SE
X[11] = FAILED SE
X[26] = GPA

X[10] <= 3.505
gini = 0.4984
samples = 802
value = [424, 378]

True   False

X[11] <= 0.1389
gini = 0.1935
samples = 258
value = [28, 230]

X[3] <= 1.1
gini = 0.3961
samples = 544
value = [396, 148]

X[26] <= 3.5164
gini = 0.0591
samples = 197
value = [6, 191]

X[26] <= 3.3676
gini = 0.4612
samples = 61
value = [22, 39]

gini = 0.0
samples = 43
value = [0, 43]

X[11] <= 0.0198
gini = 0.3313
samples = 501
value = [396, 105]

Case 1

gini = 0.0
samples = 143
value = [0, 143]

Case 2

gini = 0.1975
samples = 54
value = [6, 48]

gini = 0.0768
samples = 25
value = [1, 24]

gini = 0.4861
samples = 36
value = [21, 15]

gini = 0.4917
samples = 163
value = [92, 71]

gini = 0.1809
samples = 338
value = [304, 34]

Fig. 6. Decision tree pruned to max depth = 3

dropping out factor. A possible explanation can be giving by taking into account that *Semesters* (the amount of semesters the student has been enrolled) also has a negative coefficient: Apparently, the longer the student is enrolled the lower the risk of dropping out. Hence, since SE subject are concentrated at the end of the program, failing these classes can be a sign of non dropouts students at the end of their careers. Most important, the standard deviation of a students grades has the most positive impact in a student's dropout factor.

Table 1. Significant coefficients for logistic regression (All Semesters)

Feature	Weight
Std GPA	2.84
Failed MAT	1.30
Failed MGMT	0.71
Avg SE	−1.28
Semesters	−1.62
Failed SE	−1.99

Finally, for Watson Analytics, the service produced several charts explaining the findings. We focused in the Decision Tree shown in Fig. 8, where Watson used *GPA* as the primary variable, detecting that 99.9% students below 3.57 drop out from university. Also, having *GPA* between 3.27 and 3.55 and variable *Failed SE* been equal to and lower than 1, students will have an 81% of probability of dropout. Watson uses the variables in a different way to those found by the different models of supervised learning, however, it is evident that the findings found by Watson serve in the same way to carry out the study.

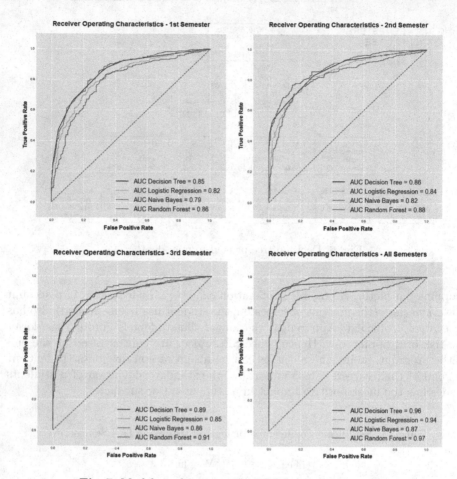

Fig. 7. Models evaluation using ROC - Area under curve

8   Deployment

This phase in CRISP–DM methodology implies to operationalize the model in the real–environment to detect risks and take decisions to prevent drop out rates. Depending on the requirements, this phase can be as simple as generating a report or as complex as implementing a repeatable data mining process. Ideally, it will be the user, not the data analyst, who will carry out the deployment steps.

However, this phase is out of scope of this work, but The findings will be shared and discussed with SE faculty in order to validate and refine the model, and implement it in a productive environment. This implementation may be deployed as a predictive web service to focus on potential dropping outs (based on prediction rules), generate early alerts and treat them properly. Afterward, the feedback of predictions and treatments should be new inputs to upgrade the model.

What is a predictive model for DROPOUT ?(Predictive strength: 83%)

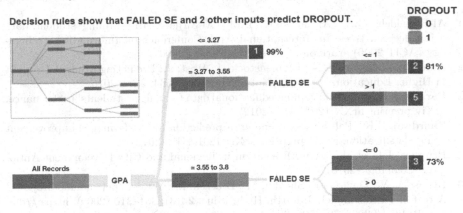

Fig. 8. Decision tree generated by Watson

9   Conclusions

Educational research has taken advantage of data mining. The current pace of applying data mining methods in this domain has increased for a variety of purposes, e.g. assessing student needs, predicting dropout rates, analyzing and improving student academic performance. Student drop out prediction is an important and challenging task.

In this paper, we showed preliminary results for predicting student attrition from a large dataset of student demographics and transcript records at different points in their degrees. In our findings, we discovered that systems engineering courses performance are correlated to physics and mathematics courses performances. The irregularity (standard deviation of term's averages) is positively correlated to dropout.

Our experimental results showed that the best AUC was achieved by random forest, from as early as the third semester of enrollment we get 0.91 AUC to last semester, where the model gave a 0.97 AUC. Four features were necessary (*Semesters, Avg SE, Failed SE, GPA*) to achieve this accuracy. It implies that courses related to SE have the greatest impact in dropout prediction.

One attractive future work is to collect a larger dataset from the whole university student database and apply the model using such data to see how it generalizes to other specific programs. In addition, other classification methods can be applied to find the most suitable method and give a better classification accuracy.

The findings must be shared and discussed with SE faculty in order to validate and refine the model and implement it in a productive environment.

References

1. Al-Radaideh, Q.A., Al-Shawakfa, E.M., Al-Najjar, M.I.: Mining student data using decision trees. In: International Arab Conference on Information Technology (ACIT 2006), Yarmouk University, Jordan (2006)
2. Aulck, L., Velagapudi, N., Blumenstock, J., West, J.: Predicting Student Dropout in Higher Education. arXiv preprint arXiv:1606.06364, June 2016
3. Baradwaj, B.K., Pal, S.: Mining educational data to analyze students' performance. arXiv preprint arXiv:1201.3417 (2012)
4. Bhardwaj, B.K., Pal, S.: Data mining: a prediction for performance improvement using classification. arXiv preprint arXiv:1201.3418 (2012)
5. Brunner, J.J., et al.: Higher Education in Regional and City Development Antioquia, Colombia (2016)
6. Brunsden, V., Davies, M., Shevlin, M., Bracken, M.: Why do he students dropout? A test of Tinto's model. J. Furth. High. Educ. **24**(3), 301–310 (2000). https://doi.org/10.1080/030987700750022244
7. Chapman, P., et al.: CRISP-DM 1.0. CRISP-DM Consortium **76**, 3 (2000)
8. Dekker, G.W., Pechenizkiy, M., Vleeshouwers, J.M.: Predicting students drop out: a case study. In: International Working Group on Educational Data Mining (2009). http://www.educationaldatamining.org/EDM2009/uploads/proceedings/dekker.pdf
9. Devasia, T., Vinushree T P, Hegde, V.: Prediction of students performance using educational data mining. In: 2016 International Conference on Data Mining and Advanced Computing (SAPIENCE), pp. 91–95. IEEE, March 2016. https://doi.org/10.1109/SAPIENCE.2016.7684167, http://ieeexplore.ieee.org/document/7684167/
10. Durso, S.D.O., Cunha, J.V.A.D.: Determinant factors for undergraduate student's dropout in an accounting studies department of a Brazilian public university. Educação em Revista **34** (2018)
11. de Educacion, M.: Spadies - sistema de prevencion y analisis a la desercion en las instituciones de educacion superior. www.mineducacion.gov.co/1621/article-156292.html. Accessed 18 July 2017
12. Jing, L.: Data mining and its applications in higher education. New Dir. Inst. Res. **2002**(113), 17–36 (2002). https://doi.org/10.1002/ir.35, https://onlinelibrary.wiley.com/doi/abs/10.1002/ir.35
13. Kim, D., Kim, S.: Sustainable education: analyzing the determinants of university student dropout by nonlinear panel data models. Sustainability **10**(4), 954 (2018)
14. Kovacic, Z.: Early prediction of student success: mining students' enrolment data. In: Proceedings of Informing Science & IT Education Conference (InSITE) (2010)
15. Márquez-Vera, C., Cano, A., Romero, C., Noaman, A.Y.M., Mousa Fardoun, H., Ventura, S.: Early dropout prediction using data mining: a case study with high school students. Expert Syst. **33**(1), 107–124 (2016). https://doi.org/10.1111/exsy.12135, http://doi.wiley.com/10.1111/exsy.12135
16. Mishra, T., Kumar, D., Gupta, S.: Mining students' data for prediction performance. In: 2014 Fourth International Conference on Advanced Computing & Communication Technologies (ACCT), pp. 255–262. IEEE (2014)
17. Romero, C., Ventura, S.: Educational data mining: a survey from 1995 to 2005. Expert Syst. Appl. **33**(1), 135 – 146 (2007). https://doi.org/10.1016/j.eswa.2006.04.005, http://www.sciencedirect.com/science/article/pii/S0957417406001266

18. Seidman, A.: Retention revisited: R= E, Id+ E & In, Iv. Coll. Univ. **71**(4), 18–20 (1996)
19. Herzog, S.: Estimating student retention and degree completion time: decision-trees and neural networks vis-á-vis regression. New Dir. Inst. Res. **2006**(131), 17–33 (2006). https://doi.org/10.1002/ir.185, https://onlinelibrary.wiley.com/doi/abs/10.1002/ir.185
20. Tekin, A.: Early prediction of students' grade point averages at graduation: a data mining approach. Eurasian J. Educ. Res. **54**, 207–226 (2014). https://eric.ed.gov/?id=EJ1057301
21. Tinto, V.: Dropout from higher education: a theoretical synthesis of recent research. Rev. Educ. Res. **45**(1), 89–125 (1975)
22. Wirth, R.: CRISP-DM: towards a standard process model for data mining. In: Proceedings of the Fourth International Conference on the Practical Application of Knowledge Discovery and Data Mining, pp. 29–39 (2000)
23. Yukselturk, E., Ozekes, S., Türel, Y.K.: Predicting dropout student: an application of data mining methods in an online education program. Eur. J. Open Distance E-Learn. **17**(1), 118–133 (2014)

Comparison of Evolutionary Algorithms for Estimation of Parameters of the Equivalent Circuit of an AC Motor

Guillermo A. Ramos[1] and Jesus A. Lopez[2]([⊠])

[1] Servicio Nacional de Aprendizaje–SENA, Cali, Colombia
garamos2@misena.edu.co
[2] Departamento de Automática y Electrónica,
Universidad Autónoma de Occidente, Cali, Colombia
jalopez@uao.edu.co

Abstract. This work shows the comparison of three evolutionary algorithms used to estimate the parameters of the equivalent circuit of a three-phase induction motor. The evolutionary algorithms utilized are Genetic Algorithm (GA), Particle Swarm Optimization (PSO) and, Bacteria Foraging Optimization (BFO). The number of executions needed to obtain confident results is calculated using statistical methods. With this value, each algorithm is used to estimate the parameters of the equivalent circuit of AC motor and, a comparison is done to select the best technique to use in a device which will estimate the efficiency of an AC motor at its operation place. The simulations show that a good selection to this application is the PSO technique.

Keywords: Evolutionary algorithms · Estimation of parameters
AC motor · Comparison of algorithms

1 Introduction

The AC motors are the machines that consume almost all the electrical energy in industrial processes. Estimation of efficiency in AC motors is a priority task to achieve an energy consumption with the least possible losses. Electric motors working with better efficiencies, use electric power in a better way and, hence, it is possible to generate a lower environmental impact in the production process.

Traditionally, to measure the efficiency of an AC motor, it should be taken to the laboratory however, it is ideal measure the efficiency at its working place [3]. For this reason, some methods have been developed that allow, by means of the estimation of parameters of the equivalent circuit of the motor, calculate its efficiency on line.

In this work we used three evolutionary algorithms to estimate the parameters of the equivalent circuit of the AC motor. The techniques that will be evaluated in this article are: Genetic Algorithm (GA) [4], PSO algorithm (Particle Swarm Optimization) [5, 10] and BFO (Bacterial Foraging Optimization) [6, 10].

The three evolutionary algorithms are compared taking as reference the results obtained in [2] and the execution time. The goal of the comparison is to choose the best

© Springer Nature Switzerland AG 2018
A. D. Orjuela-Cañón et al. (Eds.): ColCACI 2018, CCIS 833, pp. 126–136, 2019.
https://doi.org/10.1007/978-3-030-03023-0_11

algorithm to implement in a device that calculates the efficiency of the AC motor at its working place.

This document has been structured as follows: Initially, the equivalent circuit of the induction motor will be explained, then three evolutionary algorithms (GA, PSO and BFO) will be described briefly, next, it is estimated statistically how many runs are necessary for each technique to obtain confident values, with this number, the motor parameters will be estimated; the results are used to do a comparison, finally, the conclusions of the work are presented.

2 Concepts

2.1 Equivalent Circuit of the AC Motor

The equivalent circuit allows modelling the behavior of an induction motor [8]. Figure 1 shows the equivalent circuit of an AC motor.

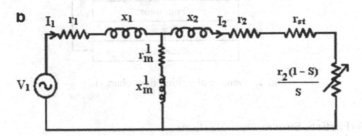

Fig. 1. Equivalent circuit of a three-phase AC motor [5]

The parameters of the equivalent circuit are:

I_1 = *Stator current*
$r_1 = r_s$ = *Stator resistance*
$x_1 = x_s$ = *Stator inductance*
r_m = *Magnetization resistance*
x_m = *Magnetization inductance*
$r_2 = r_r$ = *Rotor resistance*
$r_{st} = r_{ad}$ = *Additional resistance*
$x_2 = x_r$ = *Rotor inductance*
s = *Sliding of the rotor*

The parameters of the equivalent circuit will be the variables that evolutionary algorithms will find. With these parameters it is possible to calculate the efficiency of the motor [2]. Specifically, four parameters will be estimated using evolutionary algorithms: magnetization resistance (RM), rotor resistance (RR = R2), magnetization inductance (XM) and stator inductance (XS = X1).

2.2 Genetic Algorithm (GA)

"It is a mathematical algorithm, highly parallel that transforms a set of mathematical objects with respect to time using operations modeled according to the Darwinian principle of reproduction and survival of the fittest, and after having presented a series of genetic operations from among the which highlights sexual recombination" [11]. Figure 2 shows a flow chart of a genetic algorithm operations.

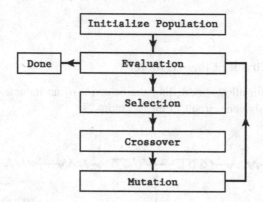

Fig. 2. Genetic algorithm flow chart [12]

2.3 PSO (Particle Swarm Optimization)

PSO is an algorithm of heuristic optimization inspired by the behavior of flocks or schools in nature. PSO allows to optimize a problem from a population of candidate solutions, denoted as "particles", moving them throughout the search space according to the modeling of the velocity and position of the particles. The modeling takes into account some control parameters as inertia, the cognitive component and the social component [9, 10].

The PSO model is expressed with two equations:

The velocity update equation

$$\vec{v}(k+1) = w\vec{v}(k) + c_1 r_{1i}(k)(P_i(k) - \vec{x}(k)) + c_2 r_{2i}(k)(g_i(k) - \vec{x}(k)) \tag{1}$$

The position update equation

$$\vec{x}(k+1) = \vec{x}(k) + \vec{v}(k+1) \tag{2}$$

Where:

$w = $ *The inertia coefficient*
$c_1 = $ *Cognitive acceleration coefficient*
$c_2 = $ *Social acceleration coefficient*
$g = $ *Better global position*
$c_1 = $ *Cognitive acceleration coefficient*
r_1 *and* $r_2 = $ *Random numbers for accelerations*
$p = $ *Current position.*

2.4   BFO (Bacteria Foraging Optimization)

BFO is a type of optimization technique based on the chemotactic behavior of bacteria (E. Coli, is one of the most studied species). Although this type of application had already been proposed in previous studies, Passino [6] included mechanisms of reproduction and dispersion of the agents improved the performance of this type of algorithms. Figure 3 shows the BFO flow chart.

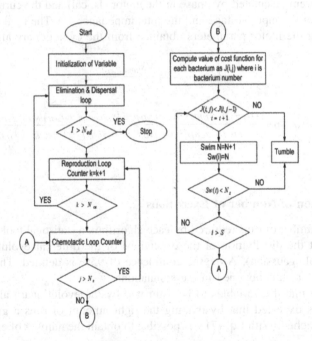

Fig. 3. BFO algorithm flow chart [10, 13]

3 Design of the Solution

3.1 Fitness Function for Parameter Estimation

The fitness function to be minimized is the following.

$$MinJ = \left|\frac{Pent_{fund,P,est}}{Pent_{fund,P,cal}} - 1\right|^2 + \left|\frac{\bar{I}s_{fund,P,est}}{\bar{I}s_{fund,P,cal}} - 1\right|^2 \tag{3}$$

This function presents the relation between the estimated power (Pent, est) with the calculated input power of AC motor (Pent, cal). The former is calculated using the motor parameters obtained from the evolutionary algorithms.

For the calculation of the estimated input power [1] equation number 4 was used

$$Pent_{fund,P,est} = 3 \cdot \left(Is_{fund,P,cal}\right)^2 \cdot \left(R_{fund,P,est}\right) \tag{4}$$

In the fitness function, there is a second term that corresponds to the relation between the current calculated by phase in the motor (Is, cal) and the current estimated (Is, est) (5) using the input voltage and the total impedance (6). The total impedance is calculated using the motor parameters obtained from the evolutionary algorithms.

$$\bar{I}s_{fund,P,est} = \frac{\bar{V}s_{fund,P,cal}}{\bar{Z}_{fund,P,est}} \tag{5}$$

$$\bar{Z}_{fund,P,est} = r_s + jx_{sFUND} + \frac{(r_m + jx_{mFUND}) \cdot \left(r_{rFUND,P} + jx_{rFUND,P} + r_{adFUND} + r_{rFUND,P} \cdot \left(\frac{1-S}{S}\right)\right)}{r_m + jx_{mFUND} + r_{rFUND,P} + jx_{rFUND,P} + r_{adFUND} + r_{rFUND,P} \cdot \left(\frac{1-S}{S}\right)} \tag{6}$$

3.2 Calculation of Number of Executions

To obtain the number of runs required by each algorithm, a statistical tool will be used. It supposes that the distribution of the executions of the different evolutionary algorithms is normal (gaussian). A degree confidence of 95% is defined. This value was verified with the results obtained in the simulations.

Considering that the variables to be estimated by the evolutionary algorithms are continuous, it is expected that by making the right amount of runs a given level of confidence is reached. With Eq. (7) it is possible to obtain the number of executions [7]

$$n = \frac{Z^2 G^2}{e^2} \tag{7}$$

Where:

n = *Sample size*
G = *standard deviation*
e = *Accuracy*
Z = *Level of confidence*

To apply the previous expression, it was performed a test of 50 runs to estimate the standard deviation of each algorithm. For this application, a confidence level of 95% (which generates a Z of 1.95) is sufficient and it corresponds to an error of $\mp 5\%$ in the final results. Experimentally it was found that with this level of confidence satisfactory results were obtained. Additionally, an accuracy of 15% is defined.

Table 1 shows the number of runs required for each parameter (XS -Stator Inductance-, RM -Magnetization Resistance-, XM -Magnetization Inductances- and RR -Rotor Resistance-) of the equivalent circuit of the AC motor using a genetic algorithm- Table 2 shows the number of runs required for PSO and Table 3 shows the number of runs required for BFO.

Table 1. Required samples for genetic algorithm (GA)

Population = 100 Crossover = 0,8			
Xs	Rm	Xm	RR
n = 280	n = 109.5458	n = 169.7098	**n = 16.2916**
Z = 1.95	Z = 1.95	Z = 1.95	Z = 1.95
G 1.6568	G 0.6482	G 1.0042	G 0.0964
e = 0.15	e = 0.15	e = 0.15	e = 0.15

Table 2. Required samples for PSO

Population = 100 C1 = 0.5 C2 = 1.2			
Xs	Rm	Xm	RR
n = 205	n = 87.2547	n = 128.4569	**n = 10.4273**
Z = 1.95	Z = 1.95	Z = 1.95	Z = 1.95
G 1.2144	G 0.5163	G 0.7601	G 0.0617
e = 0.15	e = 0.15	e = 0.15	e = 0.15

According to results, it is observed that the number of executions depends on the technique used and the parameter of the equivalent circuit estimated. The lowest number of executions is needed when we want to estimate the rotor resistance (RR) on the other hand, the highest number of executions is required when we want to estimate the stator inductance (XS). When stator inductance (XS) is estimated, the genetic algorithm is the technique that need more runs. The value required for this case is 280, so it will be the selected value to make the estimations with the three evolutionary algorithms used in this work.

Table 3. Required samples for BFO

Population = 100 Elimination Probability = 0.1			
Xs	Rm	Xm	RR
n = 234	n = 81.6777	n = 146.7258	**n = 10.2076**
Z = 1.95	Z = 1.95	Z = 1.95	Z = 1.95
G 1.3859	G 0.4833	G 0.8682	G 0.0604
e = 0.15	e = 0.15	e = 0.15	e = 0.15

4　Results

After performing the 280 executions for each evolutionary algorithm (AG, PSO and BFO), we obtain for each parameter its estimation.

Figure 4 shows the results using the three evolutionary algorithms to estimated the magnetization resistance (RM). The reference value is taken from [2].

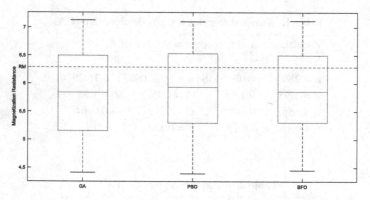

Fig. 4. Magnetization resistance (RM) comparison. Dotted red line: Reference value (Color figure online)

From the Fig. 4 it can be said that the value estimated by PSO is nearest to the ideal value of the RM. In addition, it can be mentioned that the standard deviation of the three algorithms is similar.

Figure 5 shows the results obtained using the three evolutionary algorithms to estimate the rotor resistance (RR). The reference value is taken from [2].

From the Fig. 5 it can be said that the PSO has the smallest standard deviation of the estimation of RR. In addition, the average value reached by the three techniques is similar.

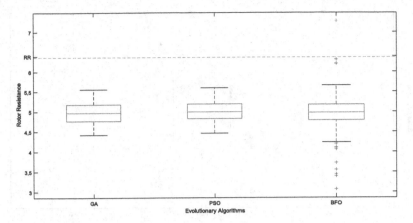

Fig. 5. Rotor resistance (RR) comparison. Dotted red line: Reference value. (Color figure online)

Figure 6 shows the results obtained using the three evolutionary algorithms to estimate the magnetization inductance (XM). The reference value is taken from [2].

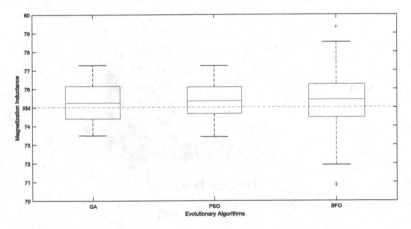

Fig. 6. Magnetization inductance (XM) comparison. Dotted red line: Reference value. (Color figure online)

From the Fig. 6 it can be said that the value estimated by GA is nearest to the ideal value of the XM. On the other hand, the PSO has a smaller standard deviation among the three algorithms.

Figure 7 shows the results obtained using the three evolutionary algorithms to estimate the stator inductance (XS). The reference value is taken from [2].

From the Fig. 7 it can be said that the estimated values by PSO and BFO are close to the ideal value of the XS. On the other hand, the PSO has a smaller standard deviation among the three algorithms.

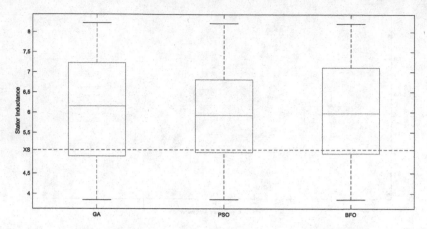

Fig. 7. Stator inductance (XS) comparison. Dotted red line: Reference value (Color figure online)

Another important information obtained from experiments is the execution time necessary to do all the runs. Figure 8 shows the execution time in seconds for the different evolutionary algorithms used in this work (GA, PSO and BFO).

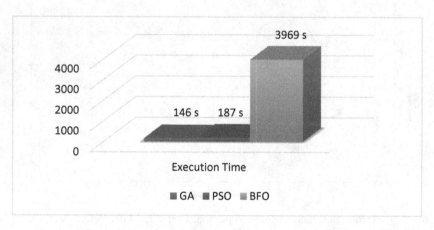

Fig. 8. Execution time in seconds of the evolutionary algorithms

Results show that the algorithm with the best performance in terms of execution time is the GA. The PSO has similar execution time but its implementation is simpler. GA can be selected if we have good computational resources but, to apply in an embedded system, it would be much more convenient to develop the application with a PSO.

5   Conclusions

- One of the questions that it is usually formulated when working on evolutionary algorithms is how many executions are necessary to be confident in the results? In this work, we answer this question using a statistical approach.
- It is necessary to clarify that the three algorithms that were compared in this document have parameters that can be adjusted to achieve better results, for this case, after finding suitable values they were left fixed.
- The estimation of the motor parameters shows that RM and XS have a big standard deviation, no matters the evolutionary algorithm used. We found that these parameters do not have a big impact in the calculation of the fitness function. On the other hand, RR and XM have parameters with a small standard deviation this mean that evolutionary algorithms realize that these parameters are import to the fitness function and they try to keep them near to the right value.
- The comparison presented in this paper is a preliminary stage to select an evolutionary computation technique to be implemented in an embedded environment, according to the results the technique selected was the PSO algorithm, since it has the best balance between accuracy, computation time and ease of implementation.

References

1. Santos, V.S.: Procedimiento para determinar la eficiencia de motores asincrónicos en presencia de desbalance y armónicos en la tensión. Tesis doctoral, Universidad Central de las Villas, Santa Clara, Cuba (2014)
2. Gómez, J.R.: Determinación de la eficiencia de los motores asincrónicos con tensiones desbalanceadas en condiciones de campo. Tesis doctoral, Universidad Central de las Villas, Santa Clara, Cuba (2006)
3. Valencia García, D.F., et al.: Estudio Del Efecto De La Distorsión Armónica De Tensión Sobre La Eficiencia Y La Potencia Del Motor Trifásico De Inducción Mediante Modelos Eléctricos Y Térmicos. Proyecto grado maestría en ingeniería énfasis en energética. Universidad autónoma de occidente (2014)
4. Gómez, J.R., Quispe, E.C., De Armas, M.A., Viego, P.R.: Estimation of induction motor efficiency in-situ under unbalanced voltages using genetic algorithms. In: 18th International Conference on Electrical Machines, ICEM 2008, pp. 1–4. IEEE (2008)
5. Sakthivel, V.P., Subramanian, S.: On-site efficiency evaluation of three-phase induction motor based on particle swarm optimization. Energy 36(3), 1713–1720 (2011)
6. Passino, K.M.: Biomimicry of bacterial foraging for distributed optimization and control. Control Syst. 22(3), 52–67 (2002)
7. Mateu, E., Casal, J.: Tamaño de la muestra. Rev. Epidem. Med. Prev. 1, 8–14 (2003)
8. Alonge, F., et al.: Parameter identification of induction motor model using genetic algorithms. IEE Proc.-Control Theory. Appl. 145, 587–593 (1998)
9. Kennedy, J.: Particle swarm optimization. In: Gass, S.I., Fu, M.C. (eds.) Encyclopedia of Machine Learning, pp. 760–766. Springer, Boston (2010). https://doi.org/10.1007/978-1-4419-1153-7_200581

10. Muñoz, M.A., López, J.A., Caicedo, E.F.: Inteligencia de enjambres: sociedades para la solución de problemas (una revisión) Ingeniería e Investigación. Universidad Nacional 2008, vol. 28, no. 2, pp. 119–130 (2008)
11. Koza, J.R.: Genetic Programing. On the Programming of Computers by Means of Natural Selection. The MIT Press, Cambridge (1992)
12. Tech Effigy Tutorials. (http://techeffigytutorials.blogspot.com.co/), http://techeffigytutorials.blogspot.com.co/2015/02/the-genetic-algorithm-explained.html. Accessed 25 May 2018
13. Song, H.M., Ibrahim, W.I., Abdullah, N.R.H.: Optimal load frequency control in single area power system using PID controller based on bacterial foraging & particle swarm optimization. ARPN J. Eng. Appl. Sci. **10**(22), 10733–10739 (2015)

Cost-Balance Setting of MapReduce and Spark-Based Architectures for SVM

Mario Alberto Giraldo Londoño[✉], John Freddy Duitama[✉], and Julián David Arias-Londoño[✉]

Universidad de Antioquia, Medellin, Colombia
mario.maat@gmail.com,
{john.duitama, julian.ariasl}@udea.edu.co

Abstract. Support Vector Machine (SVM) is a classifier widely used in machine learning because of its high generalization capacity. The sequential minimal optimization (SMO) its most popular implementation, scales somewhere between linear and quadratic in the training set size for various test problems. This fact makes using SVM to train large data sets have a high computational cost. SVM implementations on distributed systems such as MapReduce and Spark have shown efficiency to improve computational cost; this paper analyzes how data subset size and number of mapping tasks affects SVM performance on MapReduce and Spark. Also, a cost model as a useful tool for setting data subset size according to available hardware and data to be processed is proposed.

Keywords: Support vector machine · Classification · MapReduce
Spark

1 Introduction

Nowadays society presents a marked growth in the generation of large amounts of all data types, such as audio, video, images, text messages and voice messages among others. This information is generated by people and machines through social networks, e-commerce platforms and interaction with new devices such as sensors, smartphones and monitoring systems, etc. Different authors have seen great potential in the analysis of information and therefore a large number of applications in which stored data is used to determine underlying patterns in information, classify objects, detect faults and, in general, prepare predictions to support decision making have been documented in the literature [1].

The automated analysis of information is done in most cases using data mining and machine learning techniques and technologies, which were initially designed to analyze small data sets. In this regard, a lot of learning algorithms were designed to improve the modeling capacity of data, although they were not intended to process large volumes of data.

The need to analyze this data using statistical techniques and learning systems has brought new challenges in the construction of parallel data processing platforms. These platforms achieve a faster solution to complex computational issues and processing of

© Springer Nature Switzerland AG 2018
A. D. Orjuela-Cañón et al. (Eds.): ColCACI 2018, CCIS 833, pp. 137–149, 2019.
https://doi.org/10.1007/978-3-030-03023-0_12

large data volumes in terms of time. One of the current tools for managing large amounts of data is via distributed file systems such as Hadoop (HDFS), and parallel processing paradigms of information under this system: MapReduce and Spark. Consequently, many machine-learning algorithms have begun to be migrated so that they can run on Hadoop platforms using MapReduce and Spark paradigms. It improves the algorithms scalability and even there are some libraries incorporating basic machine learning techniques for use on HDFS.

Within machine learning, one of the most common issues to be solved is classification [2], which consists of designing a system able to identify which class (from a set of pre-established classes) a given object belongs to. One of the most widely used models in literature to solve classification problems is support vector machines (SVM) [2]; which were introduced in 1992 by Vapnik and Chevronenkisson. Initially they were thought to solve problems of bi-class classification using linear models; but nowadays they are used to solve multi-class classification and different regression problems.

SVM training implies the solution to an optimization problem, whose computational cost using conventional algorithms such as the interior-point can be of the quadratic order respect to number of the training data [3]. This fact turns SVMs utilization for the analysis of large amounts of data into a high computational cost, because memory requirements increase considerably affecting the training time [4]. In order to reduce this computational cost, some solutions have been proposed, such as the SMO algorithm [4], which reduces SVMs computational cost. The cost of SMO respect to training data set size is linear in memory usage; while in computation time, it scales between linear and quadratic [4]. Although SMO substantially improves SVM training time, it was not intended to work in parallel programming environments; in fact, its sequential nature makes the parallelization algorithm a complex problem to solve. In literature, different versions of SMO have been proposed to work in parallel programming environments in order to analyze large volumes of data with an improvement in training time and without affecting accuracy.

In recent years, several papers have been published [5–9] where the standard SVM is implemented under MapReduce paradigm and Spark, that is, works in which the implemented SMO algorithm is used in its standard version or through the libSVM library [10] and it is parallelized in mapping and reducing tasks. In this research, it is observed that some phases and methods in SVM parallelization are common, such as dividing data set into small subsets (randomly or through a bootstrapping technique) in the mapping phase input and assigning the number of mapping tasks according to the number of data subset. In other studies, tasks are assigned in several phases, the output of one or several mapping tasks are assigned to the input of another task. In all the papers, SMO algorithm implemented in its standard version or through the libSVM library is used. Although these works present significant improvements in training time, without substantially affecting the percentage of accuracy in classification (one or two points below the sequential algorithms) the problem of defining optimal number of partitions, or the configuration that should be used in MapReduce or Spark scheme is not addressed in order to obtain the best possible results. It is necessary to bear in mind that an application on MapReduce and Spark has limitations regarding the hardware available, since the cost of communication between the nodes or the RAM size can significantly affect the execution time of tasks or processes executed in the system [11–13].

For that reason, the present work is aimed at establishing parameters that allow relating available hardware and setting of model parameters, by means of a cost formula; which helps to define an execution architecture of SVM on MapReduce and Spark, assessing its impact on training time and on the classifier accuracy. Here, architecture is the appropriate setting of SVM execution parameters on MapReduce and Spark for a given data set and available resources.

2 Methods

This section shows the use of MapReduce and Spark to parallelize SVMs training for bi-class classification problems; likewise, a brief review of the literature onSVM and MapReduce and Spark paradigms is included.

2.1 Support Vectors Machine

Support vector machines are a supervised machine-learning model, proposed by Vapnik [2], to solve classification and regression problems. The main objective of SVM is to find a hyperplane that maximizes the margin of separation among samples of a data set.

Given a set of data of the form $(x_1, y_1), (x_2, y_2), ..., (x_n, y_n)$, where $x \varepsilon R^n$, and $y \varepsilon \{1, -1\}$ indicates the class to which the sample belongs to. The classification function of SVM would be represented by the following equation:

$$y(x) = sign\left[\sum\nolimits_{i=1}^{n} \alpha_i y_i K \langle x.x_i \rangle + b\right] \tag{1}$$

Where K is a dot product function or Kernel [3]. The coefficient α_i is obtained by solving the following convex quadratic programming problem (QP):

$$Maximize \sum\nolimits_{i=1}^{n} \alpha_i - \frac{1}{2}\sum\nolimits_{i,j=1}^{n} \alpha_i y_i \alpha_j y_j \langle x_i.x_j \rangle \tag{2}$$

Subject to:

$$\sum\nolimits_{i=1}^{n} y_i \alpha_i = 0 \quad \alpha_i \geq 0 \quad \forall\_i = 1, ..., n \tag{3}$$

Where C is a regularization parameter of the model that maintains the compensation between the accuracy and generalization of the classifier. C is a user defined parameter. x_j are called support vectors only if their corresponding $\alpha_i > 0$.

2.2   MapReduce

MapReduce is a paradigm of parallel programming for the distributed file system Hadoop [14]. A MapReduce process is executed in two phases, one of mapping and another of reduction, which facilitates the data processing in parallel.

In the mapping phase, data input is divided into fragments (splits[1]) and they are assigned to one or several tasks called sets-of-tuples (key, value) mapping. These tasks are executed in parallel and independently. Each task applies a function on each item or on the complete split and returns the ordered results again in a set of tuples (key, value). The reduction phase is also a multi-task executed in parallel which is called reducer. In this phase, the set of tuples delivered in the mapping phase are taken and those with the same key are associated to a single result. In reduction, results are also given in the form (key, value).

In general, both MapReduce input and output are stored in a distributed file system.

2.3   Spark

Spark is a framework for distributed computing intended for large-scale, linearly scalable and fault-tolerant data processing [15]. It performs memory processing using RDD data type. RDDs are a collection of partitioned and immutable objects during processing. The works in Spark are distributed in a master (master) and several slaves (slave). The master sends tasks to slaves and they return the results to the master. The types of tasks executed on Spark over the RDDs are divided into two categories: transformations and actions. The definition of these tasks is done sequentially in groups of transformations and they are executed after defining an action. This processing mode facilitates not having to reload data from the beginning in each transformation or action executed. This makes Spark very suitable for iterative algorithms requiring multiple readings on a data set, as well as for applications requiring quick queries on large data sets. This feature means that Spark is suitable for SMO algorithm execution.

3   Proposed Procedure

In order to determine which aspects of MapReduce and Spark usage affect SVM training, an implementation of the SVM model for bi-class classification was carried out using the library libSVM [10] in each of the frameworks.

SVM training is divided into two phases both in MapReduce strategy and in Spark strategy (Fig. 1): in the first phase called training, a subset of training data is taken and training is done with that data using libSVM library. In each of the trainings, a sub-model trained with a set of support vectors (SV), a set of alpha coefficients and a value of bias or b-intercept are generated. These sub-models enter the second phase called consolidation, where the support vectors and alpha coefficients are concatenated, the number of support vectors of each sub-model is summed up and the value of bias or

---

[1] Fragment of data less than or equal to a data block of Hadoop, which is processed by a mapping task. A block is a piece of data in which hadoop distributes a file in the cluster.

intercept b is calculated with a weighted mean, obtaining at the end a single trained model. Based on these two phases, four training strategies were established:

- *ST1:* training is carried out in the mapping tasks, and consolidation is carried out in the reduction task.
- *ST2:* samples are organized in a matrix structure through the mapping tasks which is sent to the reduction tasks where the training is carried out and the consolidation is carried out in a second reduction work.
- *ST3:* training is carried out during the mapping tasks. A first intermediate consolidation of several sub-models is done over combiner task in order to reduce information traffic towards the reducer and the final consolidation is carried out in the reduction task.
- *ST4:* training samples are organized in a matrix structure In the mapping tasks. The matrix is sent to the combiner tasks where the training is carried out and the consolidation phase is carried out during the reduction task.

Iterative training is carried out and the number of iterations is controlled by reaching a maximum number of iterations or because the number of support vectors did not decrease in the current iteration, keeping the same number of support vectors of the previous iteration (see Fig. 1).

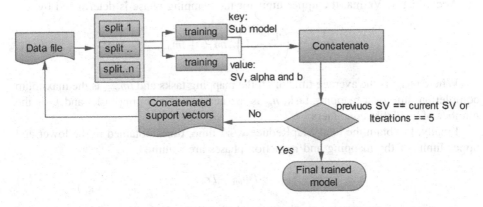

Fig. 1. Iterative SVM training on MapReduce and Spark

Figure 1 shows the training as an iterative process, where the number of support vectors decreases in each iteration. The first iteration selects a sample of about 65% of the input dataset; from second iteration, each one reduces of support vectors by about 2% with respect to its input support vectors. This behaviour impacts the model accuracy. In the first iterations accuracy ranks (increases or decreases) about 3% of the initial value. It tends to stabilize in the same way that SVM decreases after each iteration.

4 Cost Model

In [11–13], authors report that the training time decreases when the number of nodes is increased. This trend is not linear by cause of factors such as failed tasks, memory available per node and processing capabilities per node. Consequently, the objective of the cost model is to provide a tool for the computing resource management in the execution of MapReduce and Spark works, in order to optimize the performance in heterogeneous environments; in our case, optimize performance in the SVM model training execution on MapReduce and Spark.

Based on the ARIA model proposed by Verma [13] and the container concept[2] managed by the Hadoop's resource planner named Yarn (Yet Another Resource Negotiator), an application profile was created performing trainings with SVM on MapReduce and Spark with the same hardware; varying only the number of training samples for obtaining relevant information such as completion time of map tasks and reducer tasks, amount of input data, amount of data written, amount of data moved between the map and the reducer (shuffle), average time of data movement (shuffle) and processing time (CPU), among others, on MapReduce and Spark execution environments. The execution average of a map task and reduction task was calculated using that data, as well as the minimum (lower limit) and maximum (upper limit) times that the execution of an application takes, with k containers and z mapping or reduction tasks.

According to Verma, the upper limit for the mapping phase is determined by:

$$Tm_{up} = \left(\frac{n_m}{k_c} - 1\right) \cdot tm_{avg} + tm_{max} \tag{4}$$

Where tm_{avg} is the average time of all the mapping tasks and tm_{max} is the maximum completion time of a mapping task, n_m is the number of mapping tasks and k_c is the number of available containers.

Finally, to obtain the total MapReduce work time, times obtained in the lower and upper limits of the mapping and reduction phases are summed:

$$Tt_{up} = Tm_{up} + Tr_{up} \tag{5}$$

Equation (5) represents pessimistic predictions of the completion time of MapReduce work since these times are calculated for applications having a low computational cost, as in the case of a word counter [14]. In our case, the map task has a high computational cost because in this phase, training is done using a SVM model; therefore, the completion time of the work on MapReduce and Spark will depend on the completion time of the training for a determined data set, with a value C and a specific kernel function. Thus, in order to build the profile of the application as suggested by Verman, it is necessary to take into account the time SVM training takes. In

---

[2] A container represents the basic unit of resource allocation (memory and CPU) for a process execution.

the experiments conducted with the training strategy in MapReduce and Spark Fig. 1, data was taken related to the size of data set cd, expressed in bytes; the size of Split data subsets, expressed in bytes; the total training time Tt, expressed in milliseconds; the reduction time Tr, expressed in milliseconds; the consumption of CPU cores cpu, expressed in milliseconds, which represents the consolidated times in milliseconds of each mapping task. With this data, an inferential analysis is conducted to find a function that expresses the average time (tm_{avg}) of execution of a mapping task in terms of CPU consumption as a dependent variable and the split as an independent variable f $(cpu, split)$. Once this function is found, the *upper limit* (Tm_{up}) of a MapReduce work can be calculated. This function is particular for each training case due to the particularity of the training data, the value C and parameter value of kernel function used, in our case the gamma parameter g of the RBF kernel function. The average time of all mapping tasks will be given by:

$$tm_{avg} = f(cpu, split) \tag{6}$$

Where f is adjusted with data taken from the performance of initial training tests in which the split size is varied. The coicient between the total size of the data set cd and the split size will determine the number of mapping tasks:

$$n_m = cd/split \tag{7}$$

Then, the upper limit of the mapping task time would be expressed as:

$$Tm = (n_m/k_c) \times f(cpu, split) \tag{8}$$

The time of the reduction task (Tr) is calculated in the initial tests of training using a function that describes the reduction time (Tr) behavior depending on the total number of training samples cd $f(cd)$.

$$Tr = f(cd) \tag{9}$$

Thus, for our case, the total training time will be expressed as:

$$Tt = (n_m/k_c) \times f(cpu, split) + f(cd) \tag{10}$$

The Eq. (10) represents our cost model. Using this cost model we can calculate the total training time based on the available containers, training samples size, split size and *CPU* consumption. Likewise, it helps us to determine: what is the maximum split size that we should use to train x samples in a given time T?

5 Experiments

All of the experiments were run on a Hadoop cluster with one main node (NameNode) and four secondary nodes (DataNode). The main node had three cores 2.3 GHz, 23 GB RAM and 3.7 Teras on disk. Each one of the secondary nodes has two cores of 2.3 GHz, 10 GB of RAM and 3.8 Teras on disk.

Operating system Centos 6.5, Hadoop 2.6, Spark 1.5; integrated in the Cloudera version 5.4 release were installed in all of the nodes.

In the experiments the following data sets were used:

On one hand, Higgs boson[3], is a data set created from ATLAS detector event simulation at the Large Hadron Collider (LHC) at CERN. The aim of the classification problem is to differentiate the "tau tau decay" events from background noises or events. On the other hand, we have Covertype[4], data from wild areas located in the National Forest in Colorado, the objective of the classification problem here is to determine the type of forest cover. The number of instances and characteristics of each data set are detailed in Table 1.

Table 1. Data sets

Dataset	Instances	Characteristics
Boson higgs	11.000.000	21
Covertype	581.012	54

Each set of training data was divided into subsets randomly and their size was determined in bytes. A cross-validation methodology with 5 fold was used for the trainings. To keep the samples ratio of one class respect to the other and to avoid that one of the subsets was left with samples of a single class, a data pre-processing was carried out, organizing the samples by classes equally. As a result, at the end of the process was achieved a better distribution of training samples in each data subset.

In the first experiment, a set of 250 thousand samples was randomly selected from the Higgs Boson file to evaluate the training time and the precision of the strategies proposed in MapReduce and Spark, 200 thousand for training and 50 thousand for validation. Also, a search to set the parameter value C in the set {0.1, 1, 10} was done and the RBF kernel function with gamma value evaluated in the set {0.001, 0.01, 1, 10, 100} was used.

As it can be seen in Table 2, among the proposed MapReduce training strategies, the strategy with training in the mapping phase *(ST1-MR)* reports a shorter training time compared to the other proposed MapReduce training strategies. For this reason, this implementation and spark solution *(ST1-Spark)* are taken to perform the other experiments.

---

[3] Available: http://archive.ics.uci.edu/ml/datasets/HIGGS. [Accessed: 06-Feb-2016]

[4] Available: http://archive.ics.uci.edu/ml/datasets/Covertype. [Accessed: 05-Feb-2016].

Table 2. Comparative table of training strategies per training times

Strategies	Fold 1	Fold 2	Fold 3	Fold 4	Fold 5	Total training
ST1-MR	0:15:48	0:16:06	0:16:32	0:16:14	0:16:41	1:21:21
ST1-Spark	0.14:50	0:15:24	0:16:15	0:16:00	0:15:45	1:18:19

Figure 2 shows variation of performance for *ST1-MR* and *ST1-Spark* strategies using different dataset sizes. As shown, the performance is significantly different as dataset grows. Clearly, Spark implementation is more scalable, because each iteration in spark does not require write its output to disk, by contrast, MapReduce writes to disk after each iteration.

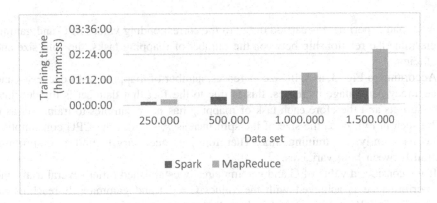

Fig. 2. Time ST1-MR and ST1-Spark

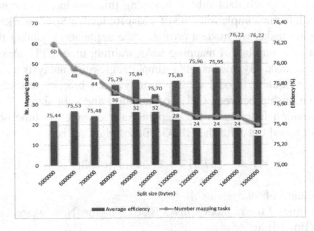

Fig. 3. Efficiency vs split size and number of mapping tasks.

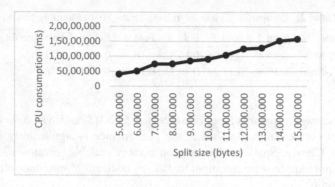

Fig. 4. CPU consumption by varying the split

A second experiment is carried out with the corresponding values of C and gamma to determine the relationship between the number of mapping tasks, the split size and the efficiency.

According to Fig. 3, as split size increases, number of mapping tasks decreases and the accuracy percentage increases, this is due to the fact that data set is divided into fewer subsets and therefore each task of mapping has more samples to train. But as it can be seen in Fig. 4, as the size of the split increases, so does the CPU consumption and, consequently, the training time. Therefore it is necessary to find a compromise solution between both variables.

If we consider a value of C and gamma already established (after several trials, the best performance is achieved with the values C = 10 and gamma = 1, reaching an accuracy of 76.03%.), Figs. 2 and 3 show that the number of mapping tasks impacts directly the training time and the classifier efficiency, as well as it determines the number of samples of each data subset. Knowing this and having a cost model that allows us to calculate the completion time of a MapReduce or Spark work, based on the number of mapping tasks, we made a profile of the application relating the following parameters: split size, number of mapping tasks, training time, CPU consumption and times in reduction task. By means of a simple regression analysis in the program Statgraphics, an x-squared model of the form $y = a + b \times x^2$ was obtained. This describes the relationship between CPU consumption and split size.

The approximation function to the mapping task completion time obtained from data of the tests carried out with the HIGGS Boson data set is:

$$y = -157.513 + 11.244,7 \times split^2 \qquad (11)$$

With an R square of 99.3% *Fig.* 4.

The average time of the reduction tasks is 561.810 ms. Now we replace Eq. 11 in the cost model (Eq. 10) as:

$$T_t = \frac{cd}{(split * k_c)} * \left(-157513 + 11244,7 * split^2\right) + T_r \qquad (12)$$

Thus, the time of training certain number of samples cd with available containers k and a split size may be determined.

Now, in order to answer the question: What is the maximum split size that we should use to train x samples in a given time?, *Eq.* 12 is taken to an optimization problem:

Objective:

$$Max f(t, split) = \frac{305,4}{(split * 8)} * \left(-157513 + 11244,7 * split^2\right) + T_r \qquad (13)$$

Subject to:

$$5 \leq split \leq 20 \quad 1 \leq t \leq 3600000$$

When solving this optimization problem, we obtain a value of 7. It corresponds to the split size in megabytes that would be needed to train a set of 305 Mb per hour with the proposed architecture (Fig 5).

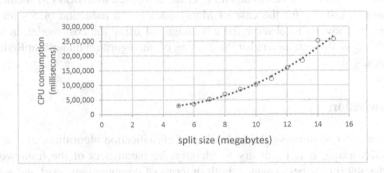

Fig. 5. Split vs CPU

Table 3 depicts the running time spent by the training process vs estimated time from the Eq. (13). It must be noted that column %error computes error between both and column named split shows different dataset sizes used. Additionally, in this proof the average error is 6.9%, the maximum error is 9.7% and the minimum error is 2.7%. This proof has as input the Higgs boson dataset.

Table 3. Training time vs estimated time using Eq. (13)

Split (mb)	# of map process	Running time	Estimated time	% error
10	8	2766357	2967318	7.3%
11	7	3124980	3357036	7.4%
12	6	2202193	2343918	6.4%
13	6	3457805	3683577	6.5%
14	6	4185656	3779439	9.7%
15	5	4478907	4599792	2.7%

Table 4 comparatively shows the result of the algorithms proposed in [9] and evaluated in the forest covertype database with the results of strategies proposed in this paper.

Table 4. Comparative table with other implementations

Algorithm	Time (s)	Accuracy
MBSSVM [16]	40	0.6913
CascadeSVM [2]	1998	0.8923
LibSVM [10]	83631	0.9615
MSM-SVM [9]	33	0.9604
ST1-MR	52	0.9620
ST1-Spark	35	0.9620

Table 4 shows that, with MapReduce and Spark strategy, efficiency increase of more than 27% was achieved in absolute terms, compared to MBSSVM method, with an increase of just 2 s for the case of MapReduce and a reduction of 5 s for Spark. These results allow us to observe the advantages of using the cost model, in order to accurately adjust the parameters for the training of the algorithm under MapReduce and Spark paradigms.

6 Conclusion

In order In order to improve the scalability of classification algorithms such as SVM, from HDFS usage, it is necessary to identify the parameters of the framework that impact the algorithms performance, both in terms of computational cost and accuracy. The research allowed to observe the effect Split size has both in the training time and in the accuracy.

The proposed model proved to be a useful tool for the configuration of the number of splits, according to available hardware and data set to be processed, and it allowed to obtain better performances in the evaluated strategies in comparison with other works from the state of the art.

References

1. Aguilar, L.J.: Big Data. Análisis De Grandes Volúmenes De Datos En Organizaciones (2013)
2. Graf, H.P., Cosatto, E., Bottou, L., Durdanovic, I., Vapnik, V.: Parallel support vector machines: the cascade SVM. In: Advances in Neural Information Processing Systems, pp. 521–528 (2005)
3. Martínez-Trinidad, J., et al.: Support vector machines for pattern classification, vol. 6256 (2010)

4. Keerthi, S., Shevade, S.K., Bhattacharyya, C., Murthy, K.R.K.: Improvements to platt's SMO algorithm for SVM classifier design. Neural Comput. **13**, 637–649 (2001)
5. You, Z., Yu, J., Zhu, L., Li, S., Wen, Z.: A MapReduce based parallel SVM for large-scale predicting protein – protein interactions. Neurocomputing **145**, 37–43 (2014)
6. Priyadarshini, A.: A map reduce based support vector machine for big data classification. Int. J. Database Theory Appl. **8**(5), 77–98 (2015)
7. Alham, N.K., Li, M., Liu, Y., Qi, M.: A MapReduce-based distributed SVM ensemble for scalable image classification and annotation. Comput. Math Appl. **66**(10), 1920–1934 (2013)
8. Çatak, F.Ö., Balaban, M.E.: A MapReduce-based distributed SVM algorithm for binary classification. Turk. J. Electr. Eng. Comput. Sci. **24**(3), 863–873 (2016)
9. Liu, C., Wu, B., Yang, Y., Guo, Z.: Multiple submodels parallel support vector machine on spark. In: 2016 IEEE International Conference on Big Data (Big Data), pp. 945–950 (2016)
10. Chang, C.-C., Lin, C.-J.: LIBSVM: a library for support vector machines. ACM Trans. Intell. Syst. Technol. **2**(3), 27:1–27:27 (2011)
11. Zhang, Z., Cherkasova, L., Loo, B.T.: Parameterizable benchmarking framework for designing a MapReduce performance model. Concurr. Comput. Pract. Exp. **26**(12), 2005–20026 (2014)
12. Chen, K., Powers, J., Guo, S., Tian, F.: CRESP: towards optimal resource provisioning for MapReduce computing in public clouds. IEEE Trans. Parallel Distrib. Syst. **25**(6), 1403–1412 (2014)
13. Verma, A., Cherkasova, L., Campbell, R.H.: ARIA: automatic resource inference and allocation for MapReduce environments. In: 8th ACM International Conference on Autonomic Computing, ICAC 2011, Karlsruhe, Germany (2011)
14. White, T.: Hadoop: The Definitive Guide, vol. 54. O'Reilly, Cambridge (2012)
15. The Apache Software Foundation, Spark Overview (2014). https://spark.apache.org/docs/latest/
16. Meng, X., Bradley, J., Yavuz, B., et al.: MLlib: machine learning in apache spark. J. Mach. Learn. Res. **17**, 1–7 (2016)

Computer Science

Nonlinear Methodologies for Climate Studies in the Peruvian Northeast Coast

Huber Nieto-Chaupis[✉]

Center of Research eHealth, Universidad de Ciencias y Humanidades,
Av. Universitaria 5175, Los Olivos, Lima39, Peru
huber.nieto@gmail.com

Abstract. We use in an explicit manner the well-known input-output methodology used in the Volterra theory to the concrete case to estimate the risks that are continuously expected due to the climatic variations in the Peruvian Northeast Coast as consequence of the arrival of phenomena such as the well-known "El Niño". We have interpreted the Volterra series as a methodological tool to calculate probabilities of risk. Thus the resulting Volterra output is therefore seen as a type of risk's probability by which a peripheral area of a large city might be affected by flooding. Under this view, the estimation of the risk depends entirely on the calculation of the parameters of the Volterra theory. The full estimation of the risk's level has used a family of input functions focused on Lorentzian and Gaussian profiles. For this end we used `Google` images by which we have focused our attention to that populations located near to rivers that are under permanent risk in summer times. This methodology can be finally seen as a scheme for disaster anticipation. We paid attention in those zones located in Tumbes city which have been affected by river overflow along the north coast of Peru in previous summer times.

Keywords: System identification · Nonlinear systems · Climate

1 Introduction

It's well known that climate sciences are seen as nonlinear phenomena and their prediction [1] is hardly done due to the stochastic and fully random behavior of climate variables, fact that makes it impossible to arrive to any robust and solid prediction in time. Although today climate sciences are mainly treated from the angle of the numerical methodologies because the modeling commonly employs nonlinear systems, certainly sustainable approximations are needed. Clearly studies of approximation for anticipating disaster aftermaths due to the arrival of the "El Niño" [2], are crucially needed because the enormous necessity that local habitants would need to know about the possible aftermaths of the arrival of this climate phenomena. In this paper we focus in the estimation of the risk's levels of certain zones geographically located in some districts which are threatened by the arrival of strong raining and that in the past have experienced

© Springer Nature Switzerland AG 2018
A. D. Orjuela-Cañón et al. (Eds.): ColCACI 2018, CCIS 833, pp. 153–164, 2018.
https://doi.org/10.1007/978-3-030-03023-0_13

disasters that have had negative effects on the aspects of education and health. For instance in the 2017 summer the Peruvian north region was victim of the strong arrival of the "El Niño" having left negative effects in the city and their districts. Essentially in this paper we focus in the estimation of risks in those peripheral areas which are located near to rivers and which might be affected to the arrival of drastic climatic changes. Thus in this paper concretely we focus in the application of a methodology based basically on input-output functions, by targeting the numerical calculation of probabilities of risk for geographical areas with a possible risk of being affected to the arrival of "El Niño". Our methodology have used convolution integration series which is perceived as a risk's probability, so that the output of the convolution is perceived as a probability. From the point of view of the systems identification theory, to apply convolution integrals is required to know both the input function and its respective kernel. Well-known examples are those such as the Wiener and Volterra series [3] that to certain extent share same mathematical structure. Normally when one faces a typical input-output problem the usage of the following equation

$$\mathcal{Y} = \hat{\mathcal{H}}\mathcal{X} \tag{1}$$

that acquires importance from the fact that it might model a system either with a linear or nonlinear behavior. Here $\mathcal{X}$ denotes the input function whereas $\hat{\mathcal{H}}$ the operator acting onto $\mathcal{X}$ producing the following effect for the case of Volterra series [4–7] for the Nth order of approximation one gets,

$$y(t) = \int \dots \int \sum_{Q=1}^{N} \prod_{j=1}^{Q} h(\tau_j) X(t - \tau_j) d\tau_j. \tag{2}$$

The Volterra series have been well used in the description of nonlinear system whose behavior in time requires a dedicate description of the modeling for input-output functions [8]. Thus, the task simply is reduced in find the Kernel that is more close to the the description of the input-output processes. However, the Volterra's series has the disadvantage of involving a huge number of kernels fact by the which one would require an enormous time consuming in order to extract the Volterra kernels that are needed to an accurate system identification. Just the precision in these type of system identification models relies entirely in the number of kernels or system parameters.

In this paper we have interpreted the output of Volterra series in terms of probability in the sense that the convolution is seen as a tool that allows to handle with probability density functions (p.d.f) that might be varying in time. In second section of this paper we present the respective mathematical machine that will be employed in the analysis of risk of geographical zones. We provide a meaning to the Volterra structure fact that keeps coherence as to the interpretation in terms o probability. Third section is reserved for results and discussion. Finally in last section the conclusion of this paper is drawn.

2   Basics of Volterra Series

2.1   The Truncated Volterra Series

The Volterra series as explicitly written in Eq. (2) have widely been used in the field of system identification inside the framework of control theory. One notable virtue of these mathematical methodologies consists in the accurate identification even being strongly nonlinear systems which are under constant perturbation and stochastic disturbs facts that in at first instance would make less accurate the identification of the system. Clearly the calculation of a huge number of kernels would involve time consumption fact can be simplified depending on the degree of approximation to be used to identify the system. It is possible to truncate the full series to a lowest degree order which might be good enough to identify the system under certain approximations. For instance, starting from the Eq. (2), a truncated second-order of Volterra series namely for $N = 2$ can be written as:

$$y(t) = \int \cdots \int \sum_{Q=1}^{N=2} \prod_{j=1}^{Q} h(\tau_j) x(t - \tau_j) d\tau_j$$

$$= y_0 + \int h(\tau_1) x(t - \tau_1) d\tau_1$$

$$+ \int \int h(\tau_1) h(\tau_2) x(t - \tau_1) x(t - \tau_2) d\tau_1 d\tau_2 \tag{3}$$

where $h(t, \tau_1)$ and $h(t, \tau_2)$ denote the Volterra kernels and $x(\tau_1)$ and $x(\tau_2)$ the input functions. Integrations are performed over the all values of τ_1 and τ_2 respectively. The quantity y_0 is seen as the error of approximation of the output in $t = t_0$. In most cases this value is unknown but normally it is accurately extracted from data. For instance, in "fast" systems input functions might be well described by the well-known Dirac-Delta functions $x(t - \tau_j) \rightarrow \delta(t - \tau_j)$ which would denote a type of initial impulse applied to the system. In "slow" systems, input functions might be well represented by continuous functions such as for example higher order polynomials such as the orthogonal polynomials.

2.2   Kernels of Volterra

A point which should be carefully stressed inside the framework of Volterra series is the methodology to be used to calculate accurately the Volterra's kernels. A well-known and effective methodology is the one by which one uses the projection of the kernel along a family of orthogonal functions. Let a set of orthogonal functions $\phi_j(\tau_j)$ then a kernel and the product of them as seen in Eq. (3) can be projected in the following form $h(\tau_1) = \sum_k c_k \phi(\tau_1)$ so for the 2th order one gets

$$h(\tau_1) h(\tau_2) = \sum_{j,k} c_j c_k \phi(\tau_j) \phi(\tau_k) = \sum_{j,k} c_{j,k} \phi(\tau_j) \phi(\tau_k). \tag{4}$$

2.3 Family of Input Functions

Because we will test the Volterra's series inside a territory of probabilities. The input functions are fully interpreted as stochastic or probabilistic distributions which are convoluted together to their kernels. Inspired on the most representative functions which have extensively used in the literature [8] such as Gaussian, Weibull and Lorentzian functions. etc. In the present analysis we focus up to two different probability distributions functions: $x \mapsto \mathcal{P}(t - \tau_j)$=Gaussian and Lorentzian functions that shall be perceived as a probability distribution function p.d.f. for our climate risk analysis. Clearly we require that these p.d.f satisfies their normalization $\int \mathcal{P}(t - \tau_j)d\tau_j = 1$. In all cases, these functions are mainly presenting a peaked behavior. Due to the structure that demands the Volterra integral. The input function would have the form $\mathcal{P}(t - \tau_j)$ thus it should be seen as a kind of probability that entirely depends on the difference $t - \tau_j$. Thus a full input $\mathcal{P}(t - \tau)$ function can be described by the sum of all p.d.f. as $\mathcal{P}(t - \tau) = p_0 \sum_\ell^Q \mu_\ell \mathcal{P}_\ell(t - \tau_j)$, where $q = 1, Q$ specifies the number of classes of p.d.f. input functions as mentioned above and characterized in some cases by having a stochastic interpretation. The parameters μ_ℓ have as role the normalization of the individual input function together to the main normalization parameter p_0. For $\xi = t - \tau_j$ we can write down that $\mathcal{P}(\xi) = p_0 \left[\mathcal{P}_G(\xi) + \mathcal{P}_W(\xi) + \mathcal{P}_L(\xi) \right]$, where $\mathcal{P}_G, \mathcal{P}_W$ and $\mathcal{P}_L$ denote for example: Gaussian, Weibull and Lorentzian probabilistic distribution functions, respectively.

2.4 Truncated Volterra Series as Probability Distribution

Considering the kernel expansion onto orthogonal polynomials we can build a Volterra series that for our ends would have to measure probabilities. Clearly it demands that the full Volterra's output would have to be normalized, thus it's required that $\int_A^B \mathcal{P}(t) = 1$. Now we turn to write down the full probability of risk containing the 2th-order truncated Volterra's series [9] in the following form

$$\mathcal{P}(t) = \int \sum_{j=1}^N c_j \phi_j(\tau_1) \mathcal{P}(t - \tau_1) d\tau_1$$

$$+ \int \sum_{j=1}^N \sum_{k=1}^M c_j c_k \phi_j(\tau_1) \phi_k(\tau_2) \mathcal{P}(t - \tau_1) \mathcal{P}(t - \tau_2) d\tau_1 d\tau_2. \tag{5}$$

The second term of Eq. (5) can be tested firstly with a input function such as a Lorentzian function for example. Thus for the the 2th order of Volterra series we have

$$\mathcal{P}(t) \approx \left| \int \sum_j^J c_j \mathcal{L}_j(\tau_2) L(t - \tau_2) d\tau_2 \right|^2, \tag{6}$$

for a lower order J, where $\mathcal{L}_j(\tau_2)$ is a Legendre polynomial, whereas the Lorentzian function is explicitly given by,

$$L(t - \tau_2) = \frac{\gamma}{1 + (t - \tau_2 - \beta)^2}. \tag{7}$$

to be used as input function inside the Volterra theory. In order to illustrate the application of these input functions we use the package Wolfram [10]. Thus we write down the normalized first order of Volterra series for two shapes of Lorentzian p.d.f., as follows

$$\mathcal{P}(t) = \int_A^B d\tau \sum_{i=q}^{4} c_q \mathcal{L}_q(\tau) P(t - \tau) \tag{8}$$

$$\mathcal{P}(t) = \int_A^B d\tau \sum_{q=1}^{4} \frac{c_q \mathcal{L}_q(\tau)}{\beta_q + (t - \tau - \lambda_1)^2}, \mathcal{P}(t) = \int_A^B d\tau \sum_{q=1}^{4} c_q \mathcal{L}_q(\tau) \mathcal{G}(t - \tau, \sigma_q) \tag{9}$$

where β_q and λ are free parameters modeling the Volterra's output.

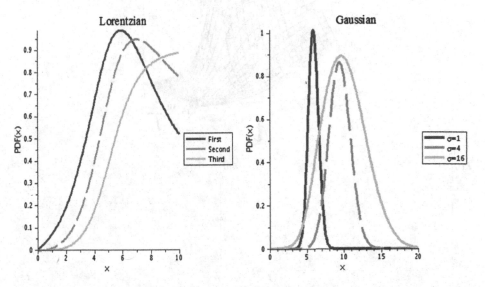

Fig. 1. Morphology of the Eq. (9) for up to three orders of the Legendre polynomials using a Lorentzian shape (left), for up to three different width of the Gaussian distributions (right). Here $t \to x$.

In Fig. 1 is demonstrated the capabilities to use the Volterra theory to be adapted inside of a framework of probabilities. We have plotted Eq. (9) (left panel) for the first three orders of the Legendre polynomials, and for three

different widths of Gaussian profiles. In Fig. 2 (Top) is plotted a 3D plot of
Eq. (9) solely for the Lorentzian case showing the consistency of the proposal
of using the truncated first order Volterra formalism. Even in the case for the
2-variables dependence of the p.d.f one can see the peaked morphology of the
curve. Therefore one expects that the integrations would follow also a peaked
distribution in a first instance. In bottom panels of Fig. 2 is plotted the integrations of Eq. (9) first term of right side corresponding to the Lorentzian case.
We used $< c_q > = 5.8910^{-5}$ (left) and 4.1910^{-5} (right), where $< c_q >$ denotes
the mean value of the parameters in according to the sum of Eq. (9). Left side
appears to conserve the peaked shape in the same manner as a pure Lorentzian
curve. In terms of probabilities the more highest probabilities appear to be inside
certain ranges along the variable t.

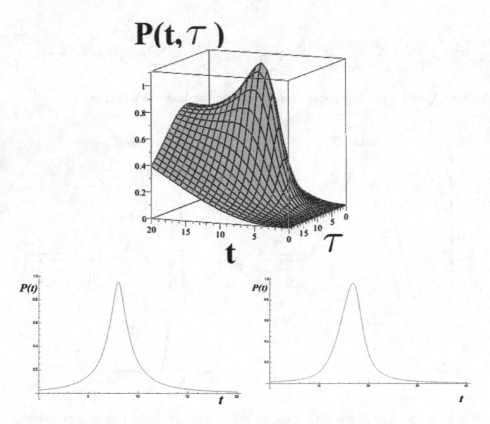

Fig. 2. (Top) 3D view of the pd.f. for the case of Lorentzian as input function. (Bottom)
Probability distribution as written in (8) and (9) for the cases where $< c_q > = 5.8910^{-5}$
(top) and 4.1910^{-5} (bottom), where $< c_q >$ denotes the mean value of the parameters
in according to the sum of Eq. 4. In both cases the width of the distribution is strongly
depending on the λ parameters.

2.5 Gaussian Function as Input

Now we turn to interpret the free parameters of the Volterra series but in a pure territory of probabilities [8,11,12]. In Fig. 3 is shown a recent (2018) `Google` map [13]. We have considered several distances between urban zones and river in order to apply the Volterra's probabilities. The central idea is to model the distribution of distances t be implemented in a full formulation of probabilities. With this in hands our next task is the calculation of the risk due to the increasing of the caudal of river in flooding times associated in a coherent manner with the full convolution integrals and the estimation of the parameters c_k, β_q and λ. Clearly the rapid increment of river caudal is directly proportional to the risk of flooding in those short distances between river and peripheral areas as indicated in Fig. 3. Another alternative to apply the Volterra theory [3] is that of the usage of Gaussian profiles that enables us to model the risk from an input function which is perceived as the risk varying with respect to time. Now we can go through the inputs functions which are written as $\mathcal{G}(t - \tau, \gamma) = g_0 \mathrm{Exp}(-\frac{t-\tau}{\gamma})^2$ with g_0 the that defines the input Gaussian function. In this case, we can formulate the full probability distribution function including first and second order, and written as

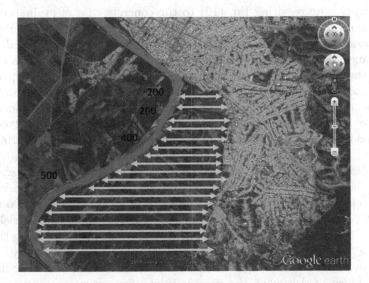

Fig. 3. Distances distribution. Satellite images taken from `Google` map [13] and the possible distances where the flooding might to reach the edges of the peripheral areas. Distances are determined in according to the map are ranging between 300 m. and 700 m.

$$\mathcal{P}(t) = \int_0^{T_1} \sum_{j=1}^{N} c_j \mathcal{L}_j(t) e^{-\beta_j t} [\mathrm{Exp}(-\frac{t-\tau_1}{\gamma^2})^2 + g(\tau_1)] d\tau_1$$

$$+ \int_0^{T_1} \int_0^{T_2} d\tau_1 d\tau_2 \sum_{j=1}^{N} \sum_{k=1}^{M} c_{j,k} \mathcal{L}_j(t) \mathcal{L}_k(t) e^{-(\beta_j+\beta_k)t}$$

$$\times [\mathrm{Exp}(-\frac{t-\tau_1}{\gamma_{j,1}^2})^2 + g(\tau_1)] \times [\mathrm{Exp}(-\frac{t-\tau_2}{\gamma_{j,2}^2})^2 + g(\tau_2)]. \tag{10}$$

It should be noted the introduction of the factors $e^{-\beta_j t}$ and $e^{-(\beta_j+\beta_k)t}$ that has as effect to decrease the probability in time. Clearly t is a variable which can be seen as days or another variable which is certainly coherent and congruent with the phenomenon under study. The second term is proportional to the square of the Legendre polynomials. Here we also introduce the random functions $g(\tau_1)$ and $g(\tau_2)$ that attempts to model the character stochastic of the flooding in time. In this work these function follow the form $r \sin \tau^2$, where r a random number.

3   Results: Application of Eq. 10

We now apply the resulting Eq. (10) to the concrete case of evaluation of the risk's level inspired in the 2017 facts of flooding in northeast coast of Peru. Our central task is to produce random probabilities and their associated distances. The distance distributions are therefore modeled by Gaussian profiles in agreement to the geometry of the geographic location between river and peripheral areas as shown in Fig. 3 whose image denotes the east part of the Tumbes city located in the left edge of Peru. During the summer epoch 2017 most of the areas near to the Tumbes river were fully flooded. In Fig. 4 are plotted up to three randomly selected curves of probability for 3 distances 50 m, 100 m and 500 m. For (A) is seen the one having high probability of risk for a distance of 50 m. However this probability decreases with respect to the days. We can see that there up to 4 days where flooding might reach this peripheral area. The curve labeled by (B) denotes the probability for a distance of 100 m. And finally, (C) for a distance of 500 m. The curve (A) indicates that the risk's level above the 50% during the first two days. The curve (B) also is showing the same case but a bit lower in their resulting probabilities. Curve (C) is fully coherent with the overflow of caudal or river. For instance the blue color curve indicates the probability of risk when the river is in overflow situation. Actually it's in a clear coherence and congruence with the rapid increasing of Tumbes river in time. This increasing is driven by the exponentials and quadratic behavior of the Legendre polynomials. The way as blue curve falls in time is interpreted as the low risk to flooding in some areas due to the reconfiguration of river as result of having flooding in other sensitive places.

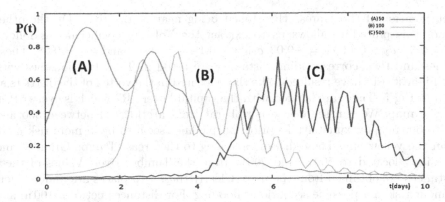

Fig. 4. Numerical integration of Eq. (10) showing up to three different aleatory selected scenarios of distances between river and peripheral areas en Tumbes city. Colors indicate the distances with the highest risks. (Color figure online)

3.1 Identification of Distances and Areas of Risk

Once the integration of Eq. (10) has been numerically tested as seen in Fig. 4 we pass to identify those areas with the highest levels of risk. According to the numerical integration of Eq. (10), $\gamma_{j,[1,2]}$ is strongly correlated to the form of the Gaussian profile in time. Although these values can be related directly to the distances, the values of c_j and c_k are estimated using the following algorithm.

- We define up to N distances with potential to be flooded from image of Fig. 3;
- We assign for each distance a Gaussian profile corresponding to the risk's level;
- Values of c_j and $c_{j,k}$ as swell as γ and β are obtained initially from a random numbers generator;
- We perform the integration of Eq. (10) to get the "crude" probability $\mathcal{P}(t)$;
- We "move" the values $\Delta\gamma_{j,[1,2]}$ and $\Delta\beta_j$; with this then:
- We estimate the shifted value $\Delta\mathcal{P}(t)$;
- If $\Delta\mathcal{P}(t) + \mathcal{P}(t) > \mathcal{P}(t)$ Then
- The deviations $\Delta\gamma$ and $\Delta\beta$ are estimated;
- If $\frac{\Delta\mathcal{P}(t)}{\mathcal{P}(t)} << \delta_i$, then:
- accept c_j and c_j, k,
- identify j-th distance
- Otherwise $j \rightarrow j+1$ and $k \rightarrow k+1$.

The computational estimation has required up to $j = 200$ that is translated as 200 potential distances to be affected by flooding. The integration was done with a partition of order of 0.05. The Romberg's method is used. In order to evaluate the proposed model, we use input function given by Gaussian profiles. Only Legendre functions up to third order were used. For this exercise of application, the resulting output is interpreted as the probability of risk in function of time. Thus, the probabilities of risk have as essential role the identification of those areas threatened being near to the river. On the other hand, this procedure allows us to extract the Volterra coefficients as follows $c_1 = 0.23, c_2 = 0.11, c_3 = -0.03, c_{11} = -0.1, c_{22} = 1.45$, and $c_{33} = 0.28$. These values and their corresponding distances are listed in Table 1. The zones with the highest risks have resulted to be those located in the vertex of the districts as seen in Fig. 5. These are labeled with the capital letter "R" which is overwritten on the map. We can see whose peripheral areas are located between 300 and up to 500 m approximately. To priori purple lines seem to be in more risk more than the yellow one. These distances belong to the areas "Pampa Grande" and "Casibo" located to 500 m and 200 m from the Tumbes river. Values of these distances are obtained from the map. Clearly these two places are under a permanent risk in a possible scenario of flooding. For distances between 100 m and 800 m. there is also a large probability of being flooded in places such as "Avau" and "Los Claveles". In Table 2 are listed the resulting places and their risk probabilities related to their distances with respect to the Tumbes river. Errors and number of parameters are also listed in Table 2.

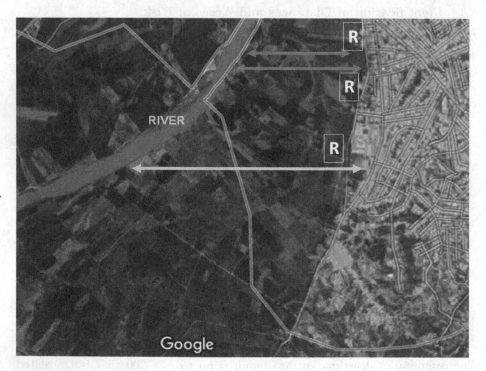

Fig. 5. Satellite images taken from Googlemap [13] and the possible distances where the flooding might to reach the edges of Tumbes city. Capital letter "R" denotes the places with the highest risk.

Table 1. Best values of Volterra coefficients and the risk probabilities $\mathcal{P}(t)$ and their associated distances.

Coefficient	0.9/50 m	0.8/100 m	0.5/300 m	0.3/200 m	0.1/100 m
c_1	0.11	0.18	0.23	0.20	0.6
c_2	−0.01	0.05	0.11	0.01	1.4
c_3	1.7	0.92	−0.03	−0.06	0.04
$c_{1,1}$	−0.6	−0.5	−0.1	−0.5	0.7
$c_{2,2}$	0.04	−1.3	1.45	1.3	0.1
$c_{3,3}$	0.01	−1.7	0.28	−1.1	0.4

Table 2. Main places with their risk probabilities and distances.

Place	Distance (km)	Risk probab	Error(%)	Number of parameters
Pampa Grande	0.5	80	10	8 ± 2
Los Claveles	0.8	70	8	10 ± 3
Avau	0.1	90	15	12 ± 4
Casibo	0.2	90	15	12 ± 4

4   Conclusion

In this paper we have applied a formalism of probability of risk based on the well-known Volterra theory to estimate the risk of certain peripheral areas inside Tumbes city located in the northeast coast of Peru, with a potential to be flooded during an eventual arrival of "El Niño". Our methodology has consisted in the usage of the truncated Volterra series which have served as a probability distribution function. With the assistance of `Google maps` we have estimated distances between river and peripheral areas that enter into a full formulation of type input-output [14,15]. Once by knowing the kernels estimated from an algorithm of identification of areas we have recognized those places located in the edges of the city as well as peripheral areas with a high risk to be flooded. This methodology can be extended to other places but more information about the geographical location of urban zones might be needed in order to decrease the error of identification.

References

1. Gorder, P.F.: Modeling El Niño: a force behind world weather. Comput. Sci. Eng. **7**(1), 5–7 (2005)
2. Takahashi, K., Martínez, A.G.: The very strong coastal El Niño in 1925 in the far-eastern Pacific. Clim. Dyn. **2017**, 1–27 (2017). https://doi.org/10.1007/s00382-017-3702-1
3. Rugh, W.J.: Nonlinear System Theory, The Volterra/Wiener Approach. Johns Hopkins University Press, Baltimore (1981)

4. Boyd, S., Chua, L.: Fading memory and the problem of approximating nonlinear operators with Volterra series. IEEE Trans. Circ. Syst. **CAS–32**(11), 1150–1161 (1985)

5. Boyd, S., Chua, L.O., Desoer, C.A.: Analytical foundations of Volterra series. J. Math. Control Inf. **1**, 243–282 (1984)

6. Boyd, S.P.: Volterra series: engineering fundamentals. Ph.D. dissertation, Department of Electrical Engineering and Computer Science, University of California, Berkeley, CA (1985)

7. Brockett, R.W.: Convergence of Volterra series on infinite intervals and bilinear approximations. In: Lakshmikanthan, V. (ed.) Nonlinear Systems and Applications, pp. 39–46. Academic, New York (1977)

8. Antzoulakos, D.: Derivation of the probability distribution functions for succession quota random variables. Ann. Inst. Stat. Math. **48**(3), 551–561 (1996)

9. Casti, J.L.: Nonlinear System Theory. Mathematics in Science and Engineering, vol. 175. Academic, Orlando (1985)

10. http://www.wolfram.com/

11. Shaikhet, L.: Stability in probability of nonlinear stochastic Volterra difference equations with continuous variable. In: Stochastic Analysis and Applications, vol. 25, no. 6, pp. 1151–1165 (2007)

12. Crouch, P.E., Collingwood, P.C.: The observation space and realizations of finite Volterra series. SIAM J. Control Optim. **25**(2), 316–333 (1987)

13. Google Maps: www.google.com/maps

14. Jing, X.J., Lang, Z.Q., Billings, S.A.: Magnitude bounds of generalized frequency response functions for nonlinear Volterra systems described by Narx model. Automatica **44**, 838–845 (2008)

15. Gilbert, E.G.: Functional expansions for the response of nonlinear differential systems. IEEE Trans. Autom. Control **AC-22**(6), 909–921 (1977)

Spectral Image Fusion for Increasing the Spatio-Spectral Resolution Through Side Information

Andrés Jerez[1] , Hans Garcia[2](✉) , and Henry Arguello[3]

[1] Department of Physics, Universidad Industrial de Santander,
Bucaramanga, Colombia
andres.jerez1@correo.uis.edu.co
[2] Department of Electrical Engineering, Universidad Industrial de Santander,
Bucaramanga, Colombia
hans.garcia@correo.uis.edu.co
[3] Department of Computer Science,
Universidad Industrial de Santander, Bucaramanga, Colombia
henarfu@uis.edu.co

Abstract. Compressive spectral imaging (CSI) allows the acquisition of the spectral information of a three-dimensional scenes by using coded projections in a sensor with lower dimension. However, the compressed sampling of information with simultaneously high spatial and high spectral resolution demands expensive high-resolution sensors. One of the main challenges in CSI is to obtain a high-quality image of high-resolution reconstructions using low-cost architectures. Single pixel camera is an approach that has had a high impact in spectroscopy, due to its low-cost implementation compared to CSI architectures with 2D sensors. On the other hand, recent works have been shown that image fusion using measurements from a CSI sensor based on side information leads to improvement in the quality of the fused image. This work proposes a spectral image fusion methodology for increasing the spatio-spectral resolution through side information and at the same time improve the reconstruction quality of the data cube with a low-cost architecture, optimizing the similarity of the reconstructed spectral image with each sensor. Simulations and experimental results for the proposed methodology show that improve the quality of the reconstruction in up to 11 dB with respect to the traditional approach of upsampling the single pixel image reconstruction through bilinear interpolation.

Keywords: Spectral imaging · Compressive sensing · Single pixel
Grayscale image · Data fusion · Side information

1 Introduction

Compressive spectral imaging (CSI) allows the acquisition of a three dimensional dataset spectral image (SI), which is composed of two spatial and one spectral

© Springer Nature Switzerland AG 2018
A. D. Orjuela-Cañón et al. (Eds.): ColCACI 2018, CCIS 833, pp. 165–176, 2018.
https://doi.org/10.1007/978-3-030-03023-0_14

dimensions. This approach increases the interpretable information that can be critical and important in different applications such as biomedical imaging for noninvasive disease diagnosis and surgical guidance [1], ground-cover classification [2], and mineral exploration and agricultural assessment in remote sensing [3]. CSI uses the principles established by compressive sensing (CS) [4] to retrieve the SI from a lower number of pixels than the traditional scanning methods, such as a wisk-broom spectrometer, push-broom or filtered camera. In CSI, the number of acquired measurements $K \in \mathbb{N}$ is considerably less than the number of reconstructed voxels, in that case the inverse problem is ill-conditioned, thus, instead of acquiring MNL samples, CSI captures $K \ll MNL$ projections of the scene, then to recover the SI $\mathbf{F} \in \mathbb{R}^{M \times N \times L}$, where L is the number of spectral bands of $M \times N$ spatial pixels, CSI assumes that the SI has a sparse representation on a given basis $\boldsymbol{\Psi}$, then a SI represented in vector form as $\mathbf{f} \in \mathbb{R}^{MNL}$ is written as a linear combination of vectors on any basis $\boldsymbol{\Psi}$, i.e., $\mathbf{f} = \boldsymbol{\Psi}\boldsymbol{\theta}$, where $\boldsymbol{\theta} \in \mathbb{R}^{MNL}$ has $S \ll MNL$ coefficients different to zero. The sensing process can be represented in matrix form as $\mathbf{g} = \mathbf{Hf}$, where $\mathbf{H}$ is the transfer function of the optical system [5] and $\mathbf{g}$ is a vector that contains the set of measurements. Hence, the inverse CSI problem consists on solving the $l_2 - l_1$ problem

$$\hat{\mathbf{f}} = \boldsymbol{\Psi} \left\{ \underset{\boldsymbol{\theta}}{\operatorname{argmin}} \|\mathbf{H}\boldsymbol{\Psi}\boldsymbol{\theta} - \mathbf{g}\|_2^2 + \tau \|\boldsymbol{\theta}\|_1 \right\}, \tag{1}$$

where $\tau \in \mathbb{R}$ is a regularization parameter.

Optical systems that employ CSI concepts often use a coded aperture to modulate the input light, which it is usually implemented by a digital micro mirror device (DMD). Furthermore, this type of architectures can employ a single pixel [6] or 2D intensity detector for acquiring the projections of the spectral scene. Some examples of architectures that include $2D$ detector are: the spatial-spectral encoded compressive spectral imager (SSCSI) [7], the coded aperture snapshot spectral imager (CASSI) [8], the dual-coded hyper-spectral imager (DCSI) [9] and the prism-mask multi-spectral video imaging system (PMVIS) [10]. It is important to remark that the implementation cost depends mainly on the resolution of the detector, in consequence, the single pixel is the lower cost architecture in these cases. Nevertheless, traditional CSI architectures are limited by the spatial resolution of the DMD.

To overcome this limitation without considerably increasing implementation cost, recent works have proposed to use high-resolution side information [11,12]. Specifically, image fusion by using measures from a CSI architecture with measures from a traditional camera, for example, the case of fusion from a CASSI architecture based on RGB side information [13], however, the implementation cost still increases due to the required $2D$ detectors. Therefore, this paper proposes a methodology for compressive data fusion based on side information by using a grayscale camera for acquiring high spatial resolution images and a single pixel camera (SPC) for retrieving the spectral information from a low-cost architecture.

This paper is organized as follows: Sect. 2 presents the mathematical description of the single pixel camera employed in this work. Then, the data fusion model for the grayscale and compressed measures is presented in Sect. 3. In order to evaluate the performance of the proposed method, several simulation results are presented in Sect. 4, employing different spectral scenes. To verify the performance of the reconstruction using side information-based architectures on real data, a testbed implementation of the system was built in the laboratory, these experimental results are included in Sect. 5. Finally, Sect. 6 presents the conclusions.

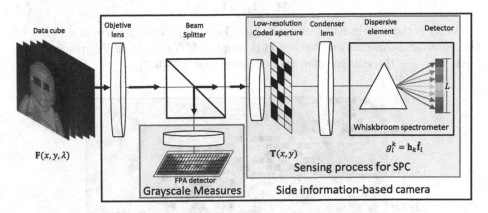

Fig. 1. Scheme of the architecture used to acquire the compressive spectral imaging based on grayscale side information.

2   SPC Sensing Model for Spectral Imaging

The single pixel architecture illustrated in Fig. 1 employs a coded aperture to modulate the input light, then, a condenser lens redirects the coded rays to a single spatial point. In this point, a single point spectrometer (Whiskbroom) is used as a detector such that all the incoming modulated light is captured in a single measurement. The sensing problem of a single spectral band $\tilde{\mathbf{f}}_l$ with low spatial resolution can be modeled as

$$g_l^k = \mathbf{h}_k\tilde{\mathbf{f}}_l, \tag{2}$$

where $l = \{1, \ldots, L\}$, $\mathbf{h}_k$ is a row vector that contains all the physical phenomena behind the architecture including the coded aperture, and k indexes each snapshot. In general, the sensing model for all shots captured for the l-th band can be written as

$$\mathbf{g}_l = \mathbf{H}\tilde{\mathbf{f}}_l, \tag{3}$$

each of these shots employs a different coded aperture pattern, $\mathbf{g}_l = \left[g_l^0, \cdots, g_l^{K-1}\right]^{\mathrm{T}}$, $\mathbf{H}$ is the sensing matrix whose rows are the vectors $\mathbf{h}_k$. Furthermore, the sensing model for all the spectral bands and K shots is given by

$$\mathbf{g} = \hat{\mathbf{H}}\mathbf{D}\mathbf{f}, \tag{4}$$

where the decimation matrix $\mathbf{D}$ represents the low spatial resolution of the SPC architecture compared with the side information that will be added to the problem in Eq. (1), $\mathbf{f}$ is the high resolution spectral image, $\mathbf{g} = \left[(\mathbf{g}_0)^{\mathrm{T}}, \cdots, (\mathbf{g}_{L-1})^{\mathrm{T}}\right]^{\mathrm{T}}$ and $\hat{\mathbf{H}}$ is the sensing matrix illustrated in Fig. 2, which can be obtained as a block diagonal matrix

$$\hat{\mathbf{H}} = \mathbf{I}_L \otimes \mathbf{H}, \tag{5}$$

where $\otimes$ is the matrix Kronecker product, $\mathbf{I}_L$ is an identity matrix of size $L \times L$, the number of columns and rows of $\hat{\mathbf{H}}$ is equal to MNL and γMNL, respectively, with $\gamma = \frac{K}{MN}$ the compression ratio. Notice that $\gamma \in [0,1]$.

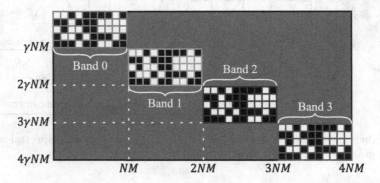

Fig. 2. SPC sensing matrix $\hat{\mathbf{H}}$, with $M = 2$, $N = 5$, $\gamma = 0.5$ and $L = 4$. White squares represent a positive value p according to the coded aperture design, black squares represent $-p$ and gray zones are 0.

For simplicity, the coded apertures patterns for the SPC architecture were random binary matrices, but recently some works are shown that is possible obtain better reconstructions with specific patterns design, this work uses coded apertures based on randomly permuted rows of the Hadamard transform with unit-norm columns proposed in [14].

3   Spectral Image Reconstruction Based on Side Information

The model used for acquiring the grayscale image assumes that the detector has an equal response for all spectral bands, then its mathematical model is represented as an identity matrix $\mathbf{I}_{N'M'}$, where $N' > N$ and $M' > M$ are the spatial dimensions of the grayscale detector for each band. The integration process of all bands can be expressed mathematically as

$$\bar{\mathbf{g}} = \sum_{l=0}^{L-1} \mathbf{I}_{N'M'}\mathbf{f}_l = \sum_{l=0}^{L-1} \mathbf{f}_l. \tag{6}$$

Equation (6) can be rewritten in matrix form as

$$\bar{\mathbf{g}} = \bar{\mathbf{H}}\mathbf{f}, \tag{7}$$

where $\bar{\mathbf{H}} = \mathbf{1}_{1 \times L} \otimes \mathbf{I}_{M'N'}$ as is shown in Fig. 3. In order to reconstruct the spatio-spectral data cube, with high spatial and high spectral resolution, it is necessary to include the measures of the SPC $\mathbf{g}$ in Eq. (4) to improve the spectral quality, and the vectorized grayscale image $\bar{\mathbf{g}}$ using the sensing model in Eq. (7) to obtain high the spatial resolution of high quality, then the new regularization problem is given by

$$\hat{\mathbf{f}} = \mathbf{\Psi} \left\{ \underset{\ddot{\boldsymbol{\theta}}}{\operatorname{argmin}} \ (1 - \tau_1) \left\| \hat{\mathbf{H}}\mathbf{D}\mathbf{\Psi}\ddot{\boldsymbol{\theta}} - \mathbf{g} \right\|_2^2 + \tau_1 \left\| \bar{\mathbf{H}}\mathbf{\Psi}\ddot{\boldsymbol{\theta}} - \bar{\mathbf{g}} \right\|_2^2 + \tau_2 \left\| \ddot{\boldsymbol{\theta}} \right\|_1 \right\}. \tag{8}$$

where $\tau_1 \in [0, 1]$ and $\tau_2 \in \mathbb{R}$ are regularization parameters. The main idea of the inclusion in the regularization problem in Eq. (8) the measure of both architectures, it is that guarantee the similitude of the spectral data cube reconstructed with both sets of measurements, this condition guide the solution near of the desired solution, and the regularization parameter τ_1 help to decide which of both sets of measurements contributes more to the solution.

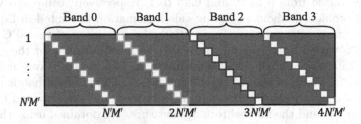

Fig. 3. Grayscale sensing matrix $\bar{\mathbf{H}}$, for $M' = 2$, $N' = 5$, and $L = 4$. White squares represent 1 and gray zones are 0.

Notice that the new compression ratio $\bar{\gamma}$ taking into account the grayscale image is given by

$$\bar{\gamma} = \frac{\gamma LMN + M'N'}{LM'N'}, \tag{9}$$

Equation (9) can be expressed in terms of the factor between the resolution of the SPC, M and N, and the resolution of the grayscale measure M' and N', this factor is calculated as $\Delta = \frac{M'}{M} = \frac{N'}{N}$, then

$$\bar{\gamma} = \frac{\gamma L + \Delta^2}{\Delta^2 L}. \tag{10}$$

In Fig. 4 is shown the analysis of compression ratio $\bar{\gamma}$ with $\gamma = 0.5$. Besides, this figure shows that for $\Delta > 1$ all values of L guarantee that the compression ratio of the SPC is holding or improved, i.e., $\bar{\gamma} \le \gamma$.

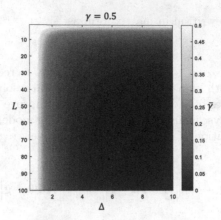

Fig. 4. Compression ratio analysis.

4   Simulations and Results

Several simulations were realized to test the performance of solving the problem from Eq. (8), when the number of bands L and rate of compression γ for the SPC measures is varied from 3 to 30 and 0.25 to 1, respectively, compared with the traditional problem without including side information presented in Eq. (1) for SPC and bilinear interpolation. In addition, the resolution of the SPC's coded aperture was varied from 16×16 to 64×64 with a resolution for the grayscale camera of 256×256 for all cases. Two different data cubes **F** were used, both with 256×256 pixels of spatial resolution and $L = 30$ spectral bands [15]. For each of these data cubes, the SPC compressive measures are obtained using the model in Eq. (4) and the monochromatic measures are obtained using the model in Eq. (7). In all cases the SI is recovered solving the minimization through the gradient projection for sparse reconstruction algorithm (GPSR) [16] with the same number of iterations. The 3D representation basis $\boldsymbol{\Psi} = \boldsymbol{\Psi}_C \otimes \boldsymbol{\Psi}_{2D}$ was used, where $\boldsymbol{\Psi}_C$ is the $1D$ discrete cosine transform and $\boldsymbol{\Psi}_{2D}$ is a $2D$ Wavelet Symmlet 8 basis. The comparisons are expressed in terms of peak signal-to-noise ratio (PSNR). All simulations were conducted and timed using an Intel Core i7-6700 @3.40 GHz processor, and 32 GB RAM memory.

4.1   Traditional Reconstruction

The grayscale images with high spatial resolution are shown in Fig. 5(a) and (b) for feathers and balloons scenes, respectively. In order to illustrate the results of the traditional methodology, the image reconstructions using the SPC measures with a low resolution coded aperture and solving the optimization problem in Eq. (1) are shown in Fig. 5(c) and (d) for each of the scenes, respectively. These results correspond to a low-resolution SI, due to the low-resolution (64×64) of the implemented coded aperture. The main idea of the proposed methodology is to combine the spatial information of the grayscale image with the spectral information acquired by the SPC to achieve an SI with high spatial-spectral resolution.

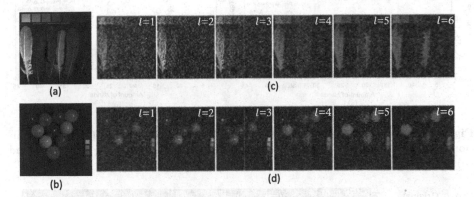

Fig. 5. The grayscale image measurement for the (a) scene 1 and (b) scene 2, with $M = N = 64$ and $M' = N' = 256$. Reconstruction of the traditional approach using the SPC measures for the (c) scene 1 and (d) scene 2.

4.2 Reconstructions Using Side Information

The reconstruction of the two data cubes using the proposed approach based on side information was simulated several times, and the average of the obtained quality in terms of the PSNR is shown in Fig. 6, when varying the amount of spectral bands for two resolution factors $\Delta = \{4, 16\}$, and three compression levels $\gamma = \{0.25, 0.5, 1\}$ for the SPC measures, these results were obtained solving the optimization problem proposed in Eq. (8) with the same number of iterations for all cases. Notice that the term traditional is referred to a low-resolution reconstruction with the original dimensions of the SPC $M \times N \times L$ using Eq. (1) and upsampled with a bilinear interpolation to $M' \times N' \times L$. Additionally, Fig. 6 shows that the proposed single pixel spectral image fusion by using a grayscale sensor as side information approach overcomes the traditional reconstruction in all cases. Specifically, in the best case, i.e., $\gamma = 1$ with $\Delta = 4$ for the scene 1 (feathers), the proposed approach provides the best reconstruction with up to 11 dB improvements over the traditional approach, and up to 8 dB of improvement for the scene 2 (balloons).

Moreover, Fig. 7 presents comparisons of the recovered spectral bands for the scene 1 and scene 2 by using traditional and proposed approach. Furthermore, the RGB mapping of the reconstructed SI is compared to the ground truth of each scene. Notice that the quality improvement of the reconstructions with the proposed methodology is evident.

4.3 Reconstruction Comparison with Respect to the Variation of τ_1

In order to analyze the behavior of the regularization parameters of the minimization problem established in Eq. (8), it is evaluated different combinations of

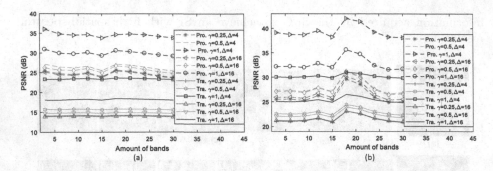

Fig. 6. Comparison of the reconstruction quality using the proposed (Pro) and traditional (Tra) approach for $\gamma = \{0.25, 0.5, 1\}$, and $\Delta = \{4, 16\}$ for the (a) scene 1 and (b) scene 2, with $M' = N' = 256$.

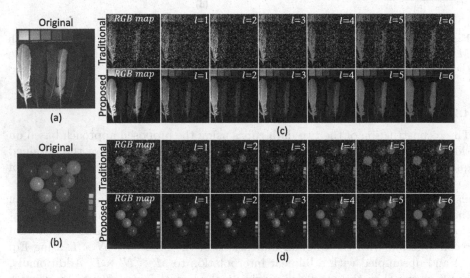

Fig. 7. Comparison between reconstructed SI using the traditional and proposed approaches for scene 1 and 2, with $M' = N' = 256$, $L = 6$ and $\gamma = 0.5$. RGB mapping of the ground truth SI for the (a) scene 1 and (d) scene 2. RGB mapping and reconstructed spectral bands for the (c) scene 1 and (d) scene 2 by using traditional and proposed approaches.

the parameter τ_1. Notice that parameter τ_1 guides the solution to more similarity with high-resolution spatial measurements acquired by the gray-scale sensor ($\tau_1 = 1$) or with low irresolution spatial resolution and spectral information acquired with SPC ($\tau_1 = 0$). Considering the compression level, the resolution factor and the number of spectral bands is analyzed the variation of the parameter τ_1. Table 1 shows the summary of PSNR results for each scenes varying the amount of spectral bands $L = \{6, 18, 30\}$, the regularization parameter $\tau_1 = \{0, 0.1, \ldots, 1\}$, the compression level $\gamma = \{0.5, 1\}$ and the resolution factor

Table 1. Summary of PSNR results using different regularization parameters for the two scenes.

		PSNR results according to amount of spectral bands										
		$\gamma = 0.5$						$\gamma = 1$				
		$\Delta = 16$			$\Delta = 64$			$\Delta = 16$			$\Delta = 64$		
Scene	L / τ_1	6	18	30	6	18	30	6	18	30	6	18	30
1	0	15.2	15.4	15.2	16.1	16.2	16.0	20.2	20.6	20.1	25.8	26.0	25.7
	0.1	16.3	16.5	16.4	25.9	26.4	25.2	21.5	22.3	21.6	35.7	35.3	34.8
	0.2	17.6	17.9	17.6	26.8	27.9	26.1	23.2	23.5	24.0	35.9	35.8	**35.1**
	0.3	19.0	19.4	18.8	27.0	28.0	26.2	25.5	24.2	24.3	**36.0**	36.0	**35.1**
	0.4	20.8	21.0	20.4	27.1	28.0	26.3	29.1	26.8	26.6	**36.0**	35.9	**35.1**
	0.5	22.7	22.7	22.0	27.3	28.1	**26.6**	28.6	30.6	28.3	**36.0**	36.0	**35.1**
	0.6	24.3	24.6	23.5	27.3	28.1	**26.6**	29.6	30.9	29.5	**36.0**	36.0	35.0
	0.7	25.2	26.0	24.7	27.3	28.1	26.5	30.1	30.7	29.5	**36.0**	**36.1**	**35.1**
	0.8	25.5	26.6	24.8	27.3	28.1	26.5	**30.2**	30.8	**29.6**	35.8	36.0	34.8
	0.9	**25.6**	**26.7**	**24.9**	**27.5**	**28.3**	26.6	**30.2**	**31.0**	29.5	**36.0**	36.0	**35.1**
	1	23.3	24.3	22.5	23.3	24.3	22.5	23.3	24.3	22.5	23.3	24.3	22.5
2	0	22.4	24.4	22.2	22.9	24.9	22.7	26.4	27.8	26.3	32.4	33.2	32.2
	0.1	23.4	25.3	23.2	27.5	31.3	27.0	28.7	30.0	28.4	39.8	42.8	39.0
	0.2	24.5	26.2	24.2	27.4	31.6	27.0	31.3	30.5	29.9	39.5	43.3	**39.5**
	0.3	25.6	27.3	25.2	27.5	31.6	27.1	31.8	34.4	31.7	39.7	43.3	39.3
	0.4	26.4	28.5	25.9	27.5	31.7	27.1	32.5	35.2	31.9	**40.0**	43.1	39.4
	0.5	26.8	29.6	26.2	27.6	31.8	27.3	31.8	35.0	**32.0**	39.3	43.4	38.8
	0.6	**26.9**	30.4	**26.3**	27.6	31.8	27.2	32.0	35.4	31.4	39.4	43.4	39.0
	0.7	26.6	30.8	26.2	27.6	31.8	27.3	32.5	35.7	31.5	39.8	43.5	39.0
	0.8	26.6	**30.9**	26.2	27.6	31.8	27.4	32.2	**36.0**	31.6	**40.0**	**43.6**	39.0
	0.9	26.6	**30.9**	**26.3**	**27.7**	**32.0**	**27.5**	**32.3**	35.8	31.8	39.7	43.4	**39.5**
	1	24.1	28.3	23.7	24.1	28.3	23.7	24.1	28.3	23.7	24.1	28.3	23.7

$\Delta = \{16, 64\}$. These results verify that the proposed fusion methodology based on Eq. (8) allows the obtainment of SI with better quality than SI reconstructed from just SPC measurements ($\tau_1 = 0$), or just grayscale measurements ($\tau_1 = 1$).

5   Experimental Setup and Results

An optical testbed implementation of the proposed architecture was constructed in the laboratory to experimentally verify its performance compared with the traditional approach. This prototype is shown in Fig. 8, and it is composed by two 100 mm objective lenses, a non-polarizing Beam-splitter to split the light, a digital micromirror device (DMD) as encoding element, a 100 mm relay lens, a F220SMA-A condenser connected through a fiber to an Ocean Optics Flame S-VIS-NIR-ES spectrometer, which is used as single pixel detector, and an AVT STINGRAY F-080B camera for acquiring the grayscale image. Measurements of a target scene were captured with this system for a compression ratio $\gamma = 1$, $M' = N' = 256$ and scale factor $\Delta = 64$.

Figure 9 shows a comparison between the reconstructed SI using the proposed and traditional approaches, specifically, in this figure is shown the spectral bands using both approaches, where the improvement is noticeable when the proposed approach is used. Here, the RGB mapping of the reconstructed scene by using the traditional and proposed approaches is compared, where it can be easily noticed that the proposed reconstruction provides a more detailed scene.

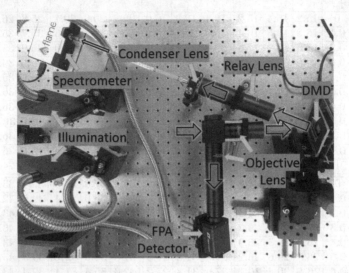

Fig. 8. Testbed implementation of the architecture used to acquire the compressive spectral measurements and side information.

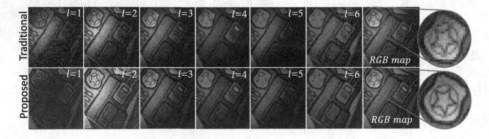

Fig. 9. Comparison between reconstructed SI using the traditional and proposed approaches for scene acquired with an optical testbed, with $M' = N' = 256$, $L = 6$ and $\gamma = 1$. Reconstructed spectral bands and RGB mapping of the SI by using traditional and proposed approaches.

6   Conclusions

A fusion methodology that allows the improvement of single pixel image reconstruction by using side information obtained from a grayscale sensor has been presented. The proposed methodology is applied to single pixel sensor, due to the fact that this architecture reduces the implementation costs. Simulation results show that the proposed methodology achieves improvements of up to 11 dB in the reconstruction quality of tested spectral images compared to the traditional approach, which uses bilinear interpolation for upsampling a low-resolution reconstruction. Finally, an optical testbed implementation of the proposed approach shows that better detailed scenes can be obtained through this system.

Acknowledgment. The authors gratefully acknowledge the optics laboratory from the High Dimensional Signal Processing (HDSP) research group for the assistance on the experimental tests. The scientific cooperation agreement subscribed between Universidad Autónoma de Bucaramanga (UNAB) and Universidad Industrial de Santander (UIS) through the summons Programa Generación ConCiencia-GEN 2017 (No. 006) for supporting this work registered under the project titled: Algoritmo de fusión de imágenes espectrales en el dominio comprimido para el aumento de la resolución espacio-espectral.

References

1. Lu, G., Fei, B.: Medical hyperspectral imaging: a review. J. Biomed. Opt. **19**(1), 10901 (2014)
2. Manolakis, D.G.: Detection algorithms for hyperspectral imaging applications. IEEE Sig. Process. Mag. **19**, 29–43 (2002)
3. Shaw, G.A., Burke, H.K.: Spectral imaging for remote sensing. Lincoln Lab. J. **14**(1), 3–28 (2003)
4. Donoho, D.L.: Compressed sensing. IEEE Trans. Inf. Theory **52**(4), 1289–1306 (2006)
5. Correa, C.V., Arguello, H., Arce, G.R.: Spatiotemporal blue noise coded aperture design for multi-shot compressive spectral imaging. J. Opt. Soc. Am. A **33**(12), 2312–2322 (2016)
6. Duarte, M., et al.: Single-pixel imaging via compressive sampling. IEEE Sig. Process. Mag. **25**(2), 1–19 (2008)
7. Lin, X., Liu, Y., Wu, J., Dai, Q.: Spatial-spectral encoded compressive hyperspectral imaging. ACM Trans. Graph. **33**(6), 233:1–233:11 (2014)
8. Wagadarikar, A., John, R., Willett, R., Brady, D.: Single disperser design for coded aperture snapshot spectral imaging. Appl. Opt. **47**(10), B44–B51 (2008)
9. Lin, X., Wetzstein, G., Liu, Y., Dai, Q.: Dual-coded compressive hyperspectral imaging. Opt. Lett. **39**, 2044–2047 (2014)
10. Cao, X., Du, H., Tong, X., Dai, Q., Lin, S.: A prism-mask system for multispectral video acquisition. IEEE Trans. Pattern Anal. Mach. Intell. **33**(12), 2423–2435 (2011)
11. Carin, L., Yuan, X., Brady, D., Tsai, T.H., Zhu, R., Llul, P.: Compressive hyperspectral imaging with side information. IEEE J. Sel. Topics Sig. Process. **9**(6), 964–976 (2015)
12. Espitia, O., Castillo, S., Arguello, H.: Compressive hyperspectral and multispectral imaging fusion. In: Proceedings of SPIE, p. 9840 (2016)
13. Galvis, L., Lau, D., Ma, X., Arguello, H., Arce, G.R.: Coded aperture design in compressive spectral imaging based on side information. Appl. Opt. **56**(22), 6332 (2017)

14. Warnell, G., Bhattacharya, S., Chellappa, R., Basar, T.: Adaptive-rate compressive sensing using side information. IEEE Trans. Image Process. **24**(11), 3846–3857 (2014)
15. Yasuma, F., Mitsunaga, T., Iso, D., Nayar, S.K.: Generalized assorted pixel camera: postcapture control of resolution, dynamic range, and spectrum. IEEE Trans. Image Process. **19**(9), 2241–2253 (2010)
16. Figueiredo, M.A.T., Nowak, R.D., Wright, S.J.: Gradient projections for sparse reconstruction: application to compressed sensing and other inverse problems. J. Sel. Topics Sig. Process. IEEE **1**(1), 586–598 (2007)

Evaluation of a Modified Current Selective Harmonic Elimination Technique Applied in Low Voltage Power Systems

Paola Beltran[1], Fabian Martínez López[1,2,3], and Oscar Flórez Cediel[3(✉)] [iD]

[1] Energy Computer System, Bogotá, Colombia
{paola.beltran, fabianemartinez}@ieee.org
[2] Easy Solutions, Bogotá, Colombia
[3] Universidad Distrital Francisco José de Caldas, Bogotá, Colombia
odflorez@udistrital.edu.co

Abstract. This document aims to present the modeling, simulation, and assessment of a modified Selective Harmonics Elimination Technique. The proposed solution could be potentially used as an Active Power Filter based on the Selective Harmonics Elimination-Pulse Width Modulation technique proposed by Enjeti [1]. An algorithm is proposed to obtain the switching angles for the Active Power Filter through numerical methods. The effectiveness of the proposed method is evaluated on simulation, and it is designed as a computer tool.

Keywords: Enjeti · Levenberg-Marquardt · Selective harmonics elimination
Trust-region-dogleg · Trust-region-reflective

1 Introduction

Harmonics are sinusoidal signals introduced in the power signal by the presence of non-linear loads, which usually affects generation, distribution and consumption users. Such harmonics include components of different frequencies that multiples the fundamental. These harmonics critically affect the power quality and the electronic equipment of the electrical systems in distribution and consumption stages [2]. Some tools have been already developed to reduce the impact of harmonics in the network by employing passive filters, active power filters (APF) or a combination thereof [3].

The inverters used in the active filters are designed in such a way that the behavior of these devices injects inverse harmonics to those generated by the load. This solution is possible through different methodologies such as pulse width modulation (PWM), neural network control, PWM with genetic algorithms, among others [4].

The SHE-PWM (Selective Harmonic Elimination) technique for control of APF has demonstrated among its most outstanding features: a high performance with a low range of switching frequencies, high voltage gain and wide bandwidth for the converters, elimination of harmonics of low order and low switching losses, through the adjusted control of harmonics and behavioral indices. Some of the investigations that

© Springer Nature Switzerland AG 2018
A. D. Orjuela-Cañón et al. (Eds.): ColCACI 2018, CCIS 833, pp. 177–186, 2019.
https://doi.org/10.1007/978-3-030-03023-0_15

apply SHE-PWM have allowed developments in cascade multilevel inverters [5–7], alternative sources of energy such as wind farms [8] and photovoltaic systems [9], active power filter (APF) [10] and even in more modern technologies such as in electric vehicles [11].

This document proposes a modification to the SHE technique for its use as APF from the SHE-PWM model proposed by Enjeti in [1]. An algorithm is developed to obtain the switching angles of the APF by numerical methods. The filter is evaluated through a simulation to validate the effectiveness of the proposal. Further developments could improve the simulation assessment, for example using real-time simulation which has proved to be useful for prototyping and model testing [12].

2 Selective Harmonic Elimination PWM

The SHE strategy starts from the definition of a power signal in the Fourier series. The switching angles that act on the inverter that generates the power signal are then obtained from the Fourier coefficients. A model for the construction of inverters by SHE-PWM is proposed in [1], starting from a quarter-wave symmetry of the power signal of the network.

$$a_n = \frac{4}{n\pi}\left[-1 - 2\sum_{k=1}^{N}(-1)^k\cos(n\alpha_k)\right] \quad and \quad b_n = 0 \tag{1}$$

Equation 1 has N variables $\alpha_1 < \alpha_2 < \alpha_3 < \ldots \alpha_N < \frac{\pi}{2}$ and a solution set which is obtained when $N - 1$. The non-linear equations that aim the $N - 1$ selective harmonics elimination of low order could be described in the following matrix system.

$$\begin{bmatrix} 2\cos(\alpha_1) & -2\cos(\alpha_2) & \cdots & 2(-1)^{(N-1)}\cos(\alpha_N) \\ 2\cos 3(\alpha_1) & -2\cos 3(\alpha_2) & \cdots & 2(-1)^{(N-1)}\cos 3(\alpha_N) \\ \vdots & \vdots & \ddots & \vdots \\ 2\cos X(\alpha_1) & -2\cos X(\alpha_2) & \cdots & 2(-1)^{(N-1)}\cos X(\alpha_N) \end{bmatrix} = \begin{bmatrix} \frac{\pi\alpha_1}{4}+1 \\ \frac{\pi\alpha_2}{4}+1 \\ \vdots \\ \frac{\pi\alpha_N}{4}+1 \end{bmatrix} \tag{2}$$

Once a solution set has been found to the matrix model of Eq. 2, the switching angles are obtained. The switching times are used to generate the trigger signals of an inverter as shown in Fig. 1. These are carried to switching times through the Eq. 3.

$$t_i = \frac{\alpha_1}{2\pi f} \tag{3}$$

where t_i is the switching time, α_1 is the switching angle and f is the fundamental frequency of the system.

The Neutral Line Signal 1 (SLN1) is the PWM signal that adjusts to quarter wave symmetry, where the switching times are distributed between 0 and T/4. This definition is constructed in four moments as shown in Fig. 2. Where $\alpha_1 < \alpha_2 < \alpha_3 < \ldots \alpha_N$ are the switching angles of matrix system in Eq. 2.

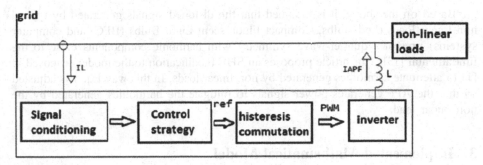

Fig. 1. APF control system blocks.

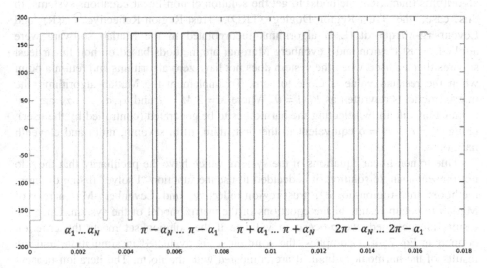

Fig. 2. SLN1

The signal neutral line 2 (SLN2) behaves in the same way as SLN1, except that it contains a phase change that helps in the elimination of the first significant harmonic. Its construction is done by intercalating the switching angles as shown in Fig. 3.

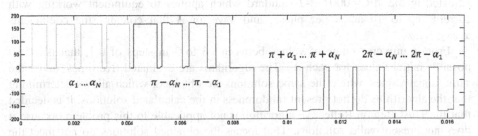

Fig. 3. SLN2

Based on the above, it is assumed that the distorted signals generated by typical non-linear loads (Led Bulbs, Compact Fluorescent Light Bulbs (BFC) and computer systems) present a quarter-wave symmetry with harmonic components close to the fundamental [13]. This article proposes an SHE modification to the model proposed in [1] to attenuate harmonics generated by non-linear loads. In this way Eq. 1 is adjusted so that the APF generates power signals to mitigate the harmonics generated by the non-linear load.

3   Implemented Mathematical Model

Given the characteristics of the matrix model presented in Eq. 2, it was necessary to use algorithms (numerical methods) to get the solution of non-linear equations systems. In this case, the Trust-Region-Dogleg (TRD), Trust-Region-Reflective (TRR) and Levenberg-Marquardt (LM) algorithms incorporated in the Matlab software were applied. Trust Region and Levenberg Marquar are methods based on non-linear least squares that are used when the system does not have zero algorithms and return a point when the residual value is close to zero. To implement the Matlab algorithms, the matrix model is rewritten as $F(x) = 0$. Where $X = 2 M - 1$ and $\alpha_1, \alpha_2, \ldots, \alpha_N$ are real values that will allow selecting the harmonics to be generated (controlled by the user); being $N = 1$ to $N = 6$ equivalent to the first, third, fifth, seventh, ninth and eleventh harmonic.

Due to non-linear equations of the system, which have the peculiarity that they do not present a single solution, it is decided to use the function "Fsolve" (using different methods: trust-region-dogleg, trust-region-reflective, and Levenberg-Marquardt) of Matlab to obtain results of the equations of the matrix model of the system. Once the results are obtained, we proceed to validate if the solution set meets the criterion defined in Eq. 2; if it complies, the solution set is evaluated in simulation, and the results of the harmonics obtained are compared with the norm. The iteration process performed to obtain the best result is presented in detail in Fig. 4.

Once the Matlab algorithm is applied to the set of Eq. 2, a validation process is done to verify if the solution set fulfills the criterion of quarter wave 3; if it complies, the solution set is evaluated in the simulation and the results of the harmonics obtained are compared with the harmonics of the load. Given Eq. 4, the following definitions are also demarcated: a vector X0 [1 × 8] that contains the initial values for α that must be adjusted to the IEC-60001 3-2 standard which applies to equipment working with currents up to 16 amps per phase, and a vector A that contains the coefficients $\alpha_1, \alpha_2, \ldots, \alpha_N$.

From a universe range of values between -3 to 3 in steps of 0.1, that is, of 60 possible initial values for each A_n. The algorithms are evaluated five times, under the same initial values, where the same solutions are obtained, which allows determining that the algorithms do not present randomness in the calculated solutions, it is decided that the Trust Region Reflective algorithm is not applicable to this problem because it does not present valid solutions. That means the obtained solutions do not meet the quarter-wave criterion. In Fig. 5 shows the results for A_1.

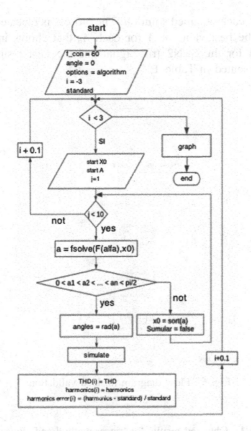

Fig. 4. Flow diagram of the algorithm to get the best value of A_n

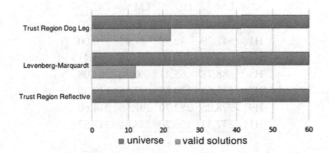

Fig. 5. Valid solutions comparison obtained by the proposed algorithm

To determine which algorithm is in the capacity to deliver the best results to achieve the attenuation of the harmonics, we proceed to observe the results obtained by the first iteration for A1 with the signal of SLN1. The error values of harmonics three, five, nine and eleven concerning the norm, the error in the fundamental, concerning the initial value of the charge current. Finally, the representation of the value of the DC component present in each obtained solution. The process is repeated for each of the A_n up to A_6 choosing the best value at A for each A_n that shows in Fig. 6. The same procedure is applied for the SLN2 trip signal and the best results obtained in the combinations are presented in Table 1.

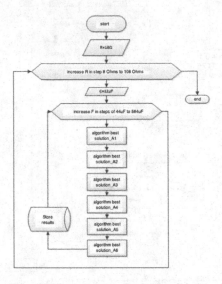

Fig. 6. Flow diagram for load validation

Table 1. Obtained results for trigger method and algorithm.

Alg	SLN1		SLN2	
	LM	DL	LM	DL
DC	18,27	1,17	0,22	0,00
H1	541,20	42,78	57,45	0,00
H3	44,97	0,00	88,78	65,41
H5	88,78	65,41	0,00	61,08
H7	135,63	130,80	97,85	117,30
H9	278,25	282,70	291,96	314,60
H11	277,52	156,20	258,81	317,53

Once this load universe has been evaluated, those that can be regulated within the framework of the IEC 61000 3-2 standard are selected. Such selection out of a total of 4.536 remaining loads a uniform sample is selected with Resistance values from 16 to 100 Ω and from Capacitance from 10 μF to 540 μF to be evaluated by the tool. This process is done by the Simulink model applying the selection algorithm described in Fig. 6.

4   Simulation Results

For the modified SHE model test, a Matlab-Simulink scheme is implemented with a simple non-linear low power single-phase load, consisting of a rectifier bridge and rectified output connected a resistor and capacitor in parallel. Establishing a set of non-linear loads that follow the RC model with a fundamental signal that complies with the characteristics of IEC 61000 3-2: fundamental current is not greater than 16 A. A simulation of 10,000 possible loads is performed in an RC combinatorial where R varies from 1 to 100 Ω in steps of 1 Ω and C from 10 μF to 1 mF in steps of 10 μF. For the 10,000 loads, the values of harmonics, total harmonic distortion, and magnitude of the fundamental are stored shows in Table 2.

Table 2. Obtained results for trigger method and algorithm.

Load non-linear	H1 (A)	H3 (A)	H5 (A)	H7 (A)	H9 (A)	H11 (A)	THDi (%)
50 Ω–525 μF	4,32	3,45	2,24	1,22	0,78	0,58	71,5
50 Ω–250 μF	4	2,78	1,47	1,08	0,86	0,44	66,6
50 Ω–25 μF	2,65	0,24	0,29	0,15	0,10	0,35	16,4
100 Ω–525 μF	2,37	1,99	1,51	0,95	0,55	0,35	75,5
100 Ω–250 μF	2,23	1,77	1,19	0,73	0,58	0,44	72,8
100 Ω–25 μF	1,39	0,27	0,22	0,07	0,07	0,06	29,4

Taking into account the time and computational cost of evaluating each of the possible loads through the model implemented in Simulink, it is decided to generate a regression from the data obtained from the simulations made to estimate the behavior of new loads and thus present an approximation to the switching angles that would allow to attenuate harmonics through of the APF. For the previous thing, the function is used regress of Matlab that generates a regression linear for each angle for the values of harmonics that the user input, as a generalization of the tool, what results in the model described in the Eq. 4.

$$\begin{bmatrix} \alpha_1 \\ \alpha_2 \\ \alpha_3 \\ \alpha_4 \\ \alpha_5 \\ \alpha_6 \\ \alpha_7 \\ \alpha_8 \end{bmatrix} = \begin{bmatrix} 1,14 & 2,81 & 3,60 & 5,10 & 5,73 & 7,87 & 8,51 & 9,89 \\ -1,72 & -4,64 & -7,61 & -11,2 & -13,4 & -23 & -24,5 & -20,59 \\ 1,13 & 2,31 & 6,84 & 15,45 & 15,9 & 14,17 & 15,5 & 21,83 \\ 1,71 & 1,87 & 3,36 & 4,32 & 5,12 & 11 & 11,52 & 8,28 \\ -0,18 & 1,72 & 2,83 & -2,48 & 0,72 & 22,6 & 23,34 & 11,43 \\ 2,01 & 4,03 & 4,47 & 8,76 & 11 & 28,77 & 31 & 25,1 \end{bmatrix} \begin{bmatrix} H_1 \\ H_3 \\ H_5 \\ H_7 \\ H_9 \\ H_{11} \end{bmatrix} \quad (4)$$

Once these values are selected, an algorithm is developed that develops the same procedure described above. The process applies the algorithm for each search for the best harmonic value according to the iteration for each position of vector A. In this way, the combination of the best initial values of A for each load is obtained. That shows in Fig. 7.

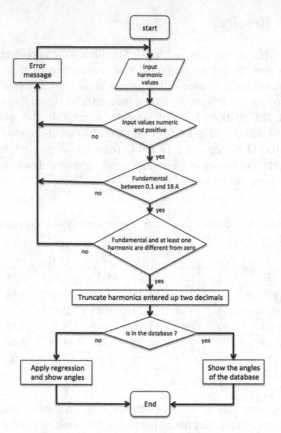

Fig. 7. Flow diagram for the user interface

The system starts from a comparison of the harmonics entered by the user against the database of simulated loads through the Simulink model. In case these values coincide with a simulated load, the interface returns the values of the angles calculated by the model. Otherwise, the system applies the regression, presenting an estimation of the switching angles to adjust the harmonics entered by the user to the parameters admitted by the IEC 61000 3-2 standard.

The design of the interface is presented in Fig. 8, which groups the data into two blocks. First, the block "Angles," where the calculated values. Second, the "Harmonics" block is shown, where harmonics are entered with positive values for the entered harmonics, positive fundamental current no higher than 16 A and different from zero and at least one harmonic (from third to eleventh) different from zero. Shows the user interface for angles obtained for load 100 Ω–525 μF.

Figure 9 shows the results obtained with the tool and the regression, compared to the maximum values allowed by the standard and the harmonics generated by the load.

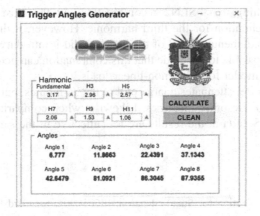

Fig. 8. User interface

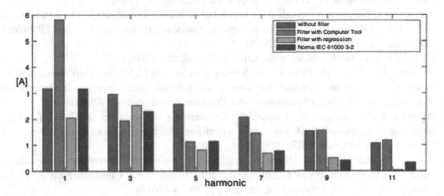

Fig. 9. Angles obtained for load 100 Ω–525 μF

5 Conclusions

The tool has a better performance with the harmonics closest to the fundamental signal, registering 79% for H3 and 34% for H5. Although the percentages are lower for other harmonics, it is emphasized that if they are attenuated. Then, it was not possible to adjust them to the norm, but they do generate a considerable improvement to the system.

Taking into account the time and computational cost of evaluating each of the possible loads through the implemented model in Simulink, it is decided to generate a regression. Such process is done from the data obtained from the simulations performed to estimate the behavior of new loads, and thus present an approximation to the switching angles that would allow attenuating harmonics through the APF.

The modification to the mathematical model proposed by Enjeti, developed throughout this work, presents an essential variation in its results when changing the algorithm to solve this equations system. The concern is left to try the modified model with other algorithms for solving systems of non-linear equations.

Due to the switching signal SLN2 in combination with the Trust Region Algorithm Dogleg presents attenuation for the third harmonic. However, it did not increase the magnitude of the fundamental signal of the load nor did it introduce DC level into the system. Then, it is possible to conclude that this combination can generate better results by generalizing the model for other non-linear loads.

The proposed model attenuates a large part of the harmonics, but its effectiveness is reduced as the harmonic order increases. Likewise, when comparing the results with the IEC 61000 3-2 Standard, the results are less effective in the same relation.

References

1. International Organization for Standardization: ISO 50001. Standard Energy Management (2011)
2. Enjeti, P., Ziogas, P., Lindsay, J.: Programmed PWM techniques to eliminate harmonics: a critical evaluation. IEEE Trans. Ind. Appl. **26**(2), 302–316 (1990)
3. Sarmiento J., Sanchez, V.: Análisis de la calidad de la energía eléctrica y estudio de carga de la Universidad Politécnica Salesiana sede Cuenca-Ecuador. Universidad Politécnica Salesiana sede Cuenca-Ecuador (2009)
4. Cydesa, Harmonics Elimination Systems. Technical data (2004)
5. Rajpriya, G., Zaidi, M.: Design and development of MATLAB Simulink based selective harmonic elimination technique for three phase voltage source inverter. In: International Conference on Advanced Computing and Communication Systems (2013)
6. Agelidis, V., Balouktsis, A., Cossar, C.: On attaining the multiple solutions of selective harmonic elimination PWM three-level waveforms through function minimization. IEEE Trans. Industr. Electron. **55**(3), 996–1004 (2008)
7. Mythili, M., Kayalvizhi, N.: Harmonic minimization in multilevel inverters using selective harmonic elimination PWM technique. In: 2013 International Conference on Renewable Energy and Sustainable Energy (ICRESE), pp. 70–74 (2013)
8. Harchegani, A., Imaneini, H.: Selective harmonic elimination pulse width modulation in single phase modular multilevel converter. In: 6th Power Electronics, Drive Systems & Technologies Conference (PEDSTC2015), 3–4 February 2015 (2015)
9. Ebadi, M., Joorabian, M.A.: New selective harmonic elimination method for wind farm using permanent magnet synchronous generator under wind speed change. In: International Aegean Conference on Electrical Machines and Power Electronics and Electromotion, Joint Conference, 8–10 September 2011 (2011)
10. Renu, V., Scholar, M.: Optimal control of selective harmonic elimination in a grid-connected single-phase PV inverter. In: 2014 International Conference on Advances in Green Energy (ICAGE), pp. 265–271, December 2014
11. Freitas, I., et al.: Singlephase active power filter for selective harmonic elimination based on synchronous frame control system. In: IEEE Applied Power Electronics Conference and Exposition - APEC, no. 5, pp. 1002–1007 (2014)
12. Wenyi, Z., Zhenhua, L., Xiaodan, M.: Electric vehicles based on selective harmonic elimination type variable voltage variable frequency speed regulation system. In: IEEE Conference and Expo Transportation Electrification Asia-Pacific (ITEC Asia-Pacific), pp. 1–5 (2014)
13. Dimas, C., Celeita, D., Ramos, G.: Simulation interface to reproduce signals with harmonic distortion of distribution systems in real-time. In: 2017 IEEE Workshop on Power Electronics and Power Quality Applications (PEPQA), Bogota, pp. 1–5 (2017)

About the Effectiveness of Teleconsults to Evaluate the Progress of Type-2 Diabetes and Depression

Huber Nieto-Chaupis[1,2](✉)

[1] Universidad de Ciencias y Humanidades, Lima, Peru
[2] Center of Research eHealth, Av. Universitaria 5175, Los Olivos, Lima39, Peru
huber.nieto@gmail.com

Abstract. We present a study using computational simulation that allows us to measure to some extent the expected impact of the teleconsults in adult patients that have been diagnosed with type-2 diabetes and which have started to exhibit depression episodes as well. Essentially we focus on the capabilities of the usage of mobile phones used to engage them to the available ehealth services offered by the public health operators. For statistical ends data is extended through Monte Carlo techniques. From the results and their respective interpretations our study have concluded that the psychological disturbs on the behavior of patients might have effect on their diabetes's treatment particularly in those living in peripheral areas of Lima city. Therefore the eHealth services might be sequentialized with psychological attentions fact that would potential the impact of the ehealth services. However we have identified that quality of service of the eHealth system might be limited seriously in those peripheral zones belonging to large cities because.

Keywords: Type-2 diabetes · Depression · Telemedicine

1 Introduction

Most of the Latin American cities have experienced in the last decades a noteworthy development in various aspects due in part to the presence of the telecommunications services. Thus, particularly in aspects related to public health, have emerged new methods that combine medicine and telecommunications, also called Telemedicine [1].

In fact, Telemedicine uses the potential of the Internet to carry out efficient consultations in real time. In fact, Telemedicine sessions are complementary to ordinary medical consultations, however, they could be crucial to perform the medical surveillance remotely and maintain also the care of the patient's health continuously.

The efficiency of a Teleconsultation would depend substantially on the operational capabilities as well as the infrastructure and equipment associated with the transmission of voice, video and data.

© Springer Nature Switzerland AG 2018
A. D. Orjuela-Cañón et al. (Eds.): ColCACI 2018, CCIS 833, pp. 187–199, 2018.
https://doi.org/10.1007/978-3-030-03023-0_16

In this paper, we present a study that deals with the probability of success of a Telemedicine session and its direct application in a specific case of public health. We have focused on the specific scenario where patients have been diagnosed with type 2 diabetes and also are passing through a season of depressive episodes.

Depress in diabetic patients might be a first cause to stop or delay an effective treatment of diabetes based on pharmacology and a strict policy based on diets.

Clearly the psychological component of the diabetes's treatment would play a crucial role as to face in a firm manner the disease and the possible ways to counteract the progress of disease.

This paper analyzes the prospective potential that the Teleconsultation could be as an important instrument that aims to extend the stability of patients [2].

We focus on patients living in sectors distant from hospital centers in the city of Lima, and whose also suffer from high glucose levels, which could lead to unexpected events [3, 4].

The lack of adequate medical attention coupled with the sudden appearance of depressive episodes further exacerbates the patient's situation that come to be very close to risks to acquire complications derived from, the disease.

Clearly the necessity to implement Teleconsultation modules [5, 6] in the vicinity of peripheral areas whose purpose is to avoid the progress of the disease in terms of stabilizing their glucose level is a must.

On the other hand, Teleconsultations also aim to decrease the number of depressive episodes per month, which would place the patient in a firm position and with conviction to continue his treatment of diabetes optimally.

This study has considered the scenario where the patient carries out a therapy based on metformin, while no other medication is administered to depression.

Our methodology has its starting point with the application of a survey that collects information about the health state of a group of diabetes patients.

The survey also measures the degree of depression in such patients and how that might lessen the patient's interest in pursuing a disciplined treatment that stabilizes blood glucose levels.

The present work has made use of computational schemes to evaluate the mathematical formulation and its implementation within a Teleconsultation system.

The paper is organized in the following way: in the second part we describe the application of a survey that collects noteworthy information of a human group located in the peripheral part of the city of Lima.

The results of the survey are used to construct a template that serves to generate extra samples. To this end we apply the Monte Carlo method to generate samples.

In third part we present a model of probabilities that is adjusted to the present study. We empathized the fact that once that the patients have surpassed the depressive episode one expects that the recover in this aspect allows them to face firmly the diabetes treatment.

The resulting statistics enters a formalism based on probabilities of success that is parameterized through a variable that encloses the quantitative specifications of a telesession, such as those characterizing an efficient Internet network.

In the fourth part we describe the operativeness of the Teleconsultation, emphasizing how the parameters of Telemedicine could be crucial for the probability of success of a Teleconsultation.

Fifth section is reserved to the results and discussion of them. Finally the conclusion of this paper is presented.

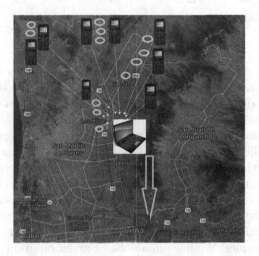

Fig. 1. Geographical location of 8 patients whom have sent their glucose's measurements information for this study. All of them belong o the north part of Lima city. The yellow arrow shows the expected connection between server and health centers located in Lima city downtown (blue squares). (Color figure online)

2   Data Analysis

We have applied a survey with a Crombach's Alpha of order of 85% to a group of patients with an old type-2 diabetes diagnosis. We have identified a pattern from the survey (official consent and ethic approbation have been done during the 2017 in the eHealth Research Center of the Universidad de Ciencias y Humanidades in Lima city) as given in Table 1 for illustration up to 6 examples.

Some examples of the patient's locations are shown in Fig. 1. Once data is scrutinized we proceed to generate up to 1000 samples with the algorithm Monte Carlo that essentially aims the simulation of additional samples from a pattern and restrictions.

Survey includes fasting glucose test, psychological test, and the global evaluation of patient in order to acquire extra information about the prevalence of other minor diseases ("other complications" see in Table 1"). Last column is

Table 1. Representative values of survey variables measured for 6 examples.

N	Age y.o.	Glucose (mg/dL)	Depress Epis monthly	Other complications	Metform.Doses g/month
6	20–30	155	1 ± 1	1	2
13	30–50	220	5 ± 2	3	10
11	50–60	270	5 ± 3	3	8
18	60–70	320	4 ± 2	4	15
2	>70	290	3 ± 2	4	12
50	45	251	3.6 ± 2	3	9.4

reserved to the information about the intake of metformin as main antidiabetic for the disease treatment. For extra sample generation we considered a data of a group of 25 type-2 diabetes patients, and we generate samples within the ranges demanded for each Monte Carlo step.

The cases real are modeled by a Gaussian function and normalized to 1. The value of the Gaussian function is evaluated inside the ranges allowed and if this number is greater than a random number then the generate sample is accepted otherwise it is rejected. With the accumulated statistics, errors are calculated. According to algorithm used, the simulation errors have been of the order of 15%. To reduce the error we need to increase the number of loops by which the algorithm is evaluated. Our algorithm has employed 7351 tries for reach the number of 1000 samples. We have made use of given program [7]. Subsequently, a curve adjustment is made. The present analysis is focuses on two variables: the value of glucose and the number of depressive episodes. Depress is identified with the Zung's questionnaire and the approbation of a psychiatry's professional. We have considered that these will be the variables of greater importance throughout the analysis and they will enter into the mathematical formulation whose probability distribution function is assumed to be a Gaussian and resulting a typical Erf function:

$$P(t_A) = \int^{t_A} dt \mathrm{Exp}[-(\frac{t}{2\sigma_F - \sigma_0})^2] = \mathrm{Erf}(t_A, 2\sigma_F - \sigma_0), \tag{1}$$

that measures the probability of being affected by diabetes and depress simultaneously. The difference given by $2\sigma_F - \sigma_0$ is perceived as the improvement or worsening of the health of patient. We consider this difference as the central value of the variable t the time by which a patient is using the eHealth services. It is important to emphasize that we have coined statistical errors to the central values in the Gaussian distributions. Clearly the characterization of both Gaussian functions is governed by the width or the amount $\sigma_F - \sigma_0$ that defines the width of the Gaussian per individual. The idea here is that the Teleconsultation has as central aim to reduce this difference which is translated in terms of efficiency. One expects that this difference falls down in time.

3   Model of Probabilities

3.1   Model Formulation

Consider the probabilities of success and fail of a certain event by which $P_S + P_F = 1$. Then $P_S = 1 - P_F$, where the P_F can be translated as the probability where there is evidence of threat or worsening with respect to the current health status of individuals. Then, when P_S is threat and perceived as a continuous function, then is possible to express the following proposition.

Proposition. The probability of success against to Q-risk situations is given by

$$P_S(\gamma) = \int \sum_{q=1}^{Q} \left[1 - P_{T_q}(x, \gamma)\right] \otimes \left[1 - P_{W_q}(x, \gamma)\right] dx \qquad (2)$$

where $P_{T_q}(x, \gamma)$ and $P_{W_q}(x, \gamma)$ the probabilities to be threated and worsened by the q-risk situation and $\mid 1 - P_{[T,W],q}(x, \gamma) \mid < 1$. These probabilities are well modeled by Eq. (1). Here γ denotes the quality of service of the teleconsultation.

Proof. Let the fail probability of success against to any disease which can be worse, then

$$P_F = P_{F,T_1} \otimes P_{F,W_1} + \ldots + P_{F,T_{Q-1}} \otimes P_{F,W_{Q-1}} + P_{F,T_Q} \otimes P_{F,W_Q}, \qquad (3)$$

and

$$P_S = P_{S,T_1} \otimes P_{S,W_1} + \ldots + P_{S,T_{Q-1}} \otimes P_{S,W_{Q-1}} + P_{S,T_Q} \otimes P_{S,W_Q} \qquad (4)$$

for Q-risk situations, then

$$P_S = [1 - P_{F,T_1}] \otimes [1 - P_{F,W_1}] + \ldots + [1 - P_{F,T_{Q-1}}] \otimes [1 - P_{F,W_{Q-1}}]$$

$$+ [1 - P_{F,T_Q}] \otimes [1 - P_{F,W_Q}] = \sum_{\ell}^{Q} [1 - P_{F,T_\ell}] \otimes [1 - P_{F,W_\ell}]. \qquad (5)$$

Now for the case where P_{F,T_ℓ} and P_{F,W_ℓ} are probability densities, then

$$P_S = \int dx \sum_{\ell}^{Q} [1 - P_{F,T_\ell}(x)] \otimes [1 - P_{F,W_\ell}(x)] \qquad (6)$$

is a number.

   Under the assumption that the P_{F,T_ℓ} and P_{F,W_ℓ} are depending upon a free parameter namely $\gamma \in \Re$ subsequently $P_S(\gamma) = \int dx \sum_{\ell}^{Q} [1 - P_{F,T_\ell}(x, \gamma)] \otimes [1 - P_{F,W_\ell}(x, \gamma)]$ is now a continuous function depending on γ.

In praxis, one is interested in those cases where there is at least one risk situation per individual. Thus for the cases of diabetes and other disease such as depress, Eq. (6) can be written as follows

$$P_{S1}(\gamma) = \int \eta_1(x,\gamma)\,[1 - P_{\mathrm{T}}(x,\gamma)]\,[1 - P_{\mathrm{W}}(x,\gamma)]\,dx \tag{7}$$

$$P_{S2}(\gamma) = \int \eta_2(x,\gamma)\,[1 - P_{\mathrm{T}}(x,\gamma)]\,[1 - P_{\mathrm{W}}(x,\gamma)]\,dx \tag{8}$$

$$P_{S3}(\gamma) = \int \eta_3(x,\gamma)\,[1 - P_{\mathrm{T}}(x,\gamma)]\,[1 - P_{\mathrm{W}}(x,\gamma)]\,dx, \tag{9}$$

where the $\eta_{[1,2,3]}(x,\gamma)$ are functions whose role is that of keeping the resulting integration between 0 and 1. All these integrations are performed over the variable x denoting the age of individuals. Here we understood P_{S1}, P_{S2} and P_{S3} as the probabilities for diabetes, depress and another extra disease respectively that is identified by the doctor or eHealth system.

4   Running a Teleconsult

4.1   Usage of the Success's Probability

Assuming that the parameters associated with a Teleconsultation are directly related to those as are commonly used in telecommunications.

We will assign values to up to four parameters in order to determine the efficiency of Teleconsultation [7]. Normally one expects to use only parameters directly linked to telecommunications, however we will implement a parameter that is closely related to the interaction between the health specialist and the patient and that will be given by the effective number of Teleconsultation per day, week or month. Table 2 summarizes the values that can be used for Teleconsultation parameters and their associated efficiency, which we will denote by ρ.

4.2   Variables of the Teleconsultation

In Table 2 are listed the main variables of the Teleconsultation. The more critic one turns out to be the phone call voice call that simply can do either the health specialist or patient. One expects that this can be done one or twice per day. Clearly one expects that at least one doctor is available to attend the call. According to the experience a voice call of 5 min might be enough to perceive the status of patient. In terms of the telecommunications parameters, the Quality of Service would be of order of 50% that means that at least one of two calls were successful.

Another variable of importance is the video calls but strongly restricted due to the costs that it demands. Taking into account the social layer of origin of

Table 2. Variables that define the Teleconsultation.

Variable	Value	Doctor available	Duration telecon	QoS (%)
Phone calls	1–2/day	1–2	5 min	50
Video calls	1/day	1	2.5 min	3
Message	3/day	1	1	75
eSurvey	2/week	1	10 min	75
Data server communication	1/day	0	10 min	80

the patients one expects that this variable might be inhibited of the circuit of Teleconsultation as expected from Fig. 2.

However the Teleconsultation might be entirely successful when text messages are used as a tool to request information about the ongoing activities of patients fact that enrich the landscape of doctor to make much more precise the next Teleconsultation.

Online survey might be applied as another strategy for extracting information of patiens when most of them are under an unexpected silent per days and weeks. Here the Psychologist would play a role crucial in the sense of the identification of those patients which might exhibit an anomalous behavior fact that can be perceived as a certain abandon of the anti-diabetic pharmacology.

Finally the dynamics of a data server receiving and processing data all days turns out to be fundamental to a optimal circuit of Teleconsultation.

4.3   The Main Parameter of a Teleconsultation

For this study we define this Teleconsultation parameter as follows

$$\rho = \frac{N_T}{N_P + N_D}, \tag{10}$$

where N_T, N_P and N_D denote the number of Teleconsultations, patients and doctors or psychologists available to the Teleconsultation. If ρ enters as an independent variable, so for large values the probability of failure decreases while success increases. This can only be possible if the disturbing functions $P_W(\rho, \mu)$ is strongly oscillating with a random argument namely μ. A function that has been tested for this. To apply the probability model of success of the Teleconsultation, we will propose the scheme as shown in Fig. 2. Consider the j-patient who applies a glucose test. The patient sends the result to a server by means of a text message. The server evaluates Eq. (6). (i) The result is stored for the glucose test history of the diabetic patient, (ii) if the probability of success is small, then a short period prediction is estimated. In parallel, a text message is sent to one of the following specialists: doctor, nurse or psychologist. On the other hand, the number of depressive episodes are counted from the beginning of the usage of the eHealth services. It is assumed that all telesesions are stored on the server [8]. If the probability of success is greater to any parameter proposed as a set-point, then only the telesession is saved and the software continues receiving messages from the other patients. To illustrate the procedure consider for example if $P_S = 0.15$ then the server uses Eq. (6) to approximate the Erf functions to the maximum values of failure probabilities. In this case are $P_F = 0.75$ to counteract an increase in glucose and stop an increase in the number of depressive episodes. This means that depression could be a priority to face a spontaneous increasing of glucose in the patient.

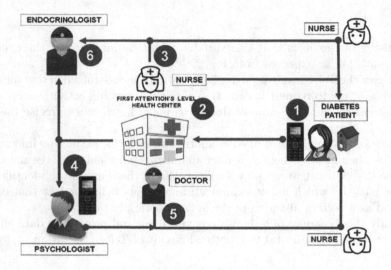

Fig. 2. Design of a scheme of Teleconsultation based on the permanent sending of glucose values (fast glucose test) and continuous monitoring of psychological episodes as well.

5   Results and Discussion

To evaluate the efficiency of the teleconsultation proposal as shown in Fig. 2, we proceed to calculate the prediction over a short period by which we consider that the efficiency has an intermediate value of 0.5, which is the scenario that best fits a peripheral area in Lima city. Our methodology has considered the results after of having implemented the teleconsultation scheme from the 5th day. Specifically, we have aimed to register those patients with their glucose values for the next two weeks or approximately 15 days. We have taken the 1000 Monte Carlo samples previously simulated, and we have evaluated it individually with the following value $\sigma_F - \sigma_0 = 14.5$. The integrals have been evaluated numerically with the Romberg method following the algorithm of [7]. Then, we have compared it

Teleconsultations Simulation

```
1    DO K = 1, K
2    γ_k, t_k, Δ
3    t_k = 0.01 + Δ
4    DO J = 1,1000 MC steps
5    P_T(η(t_k), Δt_k, γ_k)
6    P_W(η(t_k), Δt_k, γ_k) Eq. 6
7    P_S = 1 - P_F
8    CALL RANDOM R_1
9    CALL RANDOM R_2
10   IF P_S > R_1
11   THEN N_T = N_T + 1 Available Teleconsultations
12   THEN N_P = N_P + 1 Number of Patients
13      IF P_S > 0.95
14      THEN N_D = N_D + 1 Available Doctors
15      ENDIF
16   QoS: ρ = N_T/(N_P+N_D)
17   THEN N_S = N_S + 1
18      IF P_F > R_2
19      THEN N_F = N_F + 1
20      FILL HISTOGRAMS P_S(ρ), P_S(t_k)
21      ENDIF
22   ENDIF
23   ENDIF
24      ERROR = ΔP_S/P_S
25      ERROR = ΔP_F/P_F
26   ENDDO
27   ENDDO
28   K→ K+1
```

with a random number generated by a probability distribution function that has follows the morphology of a Weibull distribution. In the top panel of Fig. 3, the results of the success probability using the teleconsultation has shown to be promising for high values of γ fact that is translated as a substantial difference between the number of attended teleconsults and the patients requesting one of them. On the other hand, it also can be understood as the success to face diabetes but with depress in a minor extent.

5.1   Limitations of the Teleconsult

Interestingly a scenario to be discussed is plotted in bottom panel of Fig. 3 where the resulting integrations acquire the form of step function. One can see that these success probabilities are decaying with respect to time fact that is

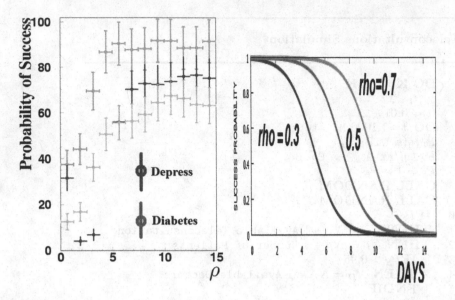

Fig. 3. Left: the success of the teleconsult as function of the γ parameter that measures the quality of the teleconsult. Green error bars denote the error per bin. Right: the success probability for the recovery of patient as function of days after the initialization of the teleconsult, for 3 different values of ρ. While this number is small there is a fail of the teleconsult to be effective in the short term. (Color figure online)

perceived as a strong limitation of the teleconsult. Thus teleconsults are the sum of psychological therapies for the treatment of depression, and on the other hand, phone calls with the Endocrinologist or a nurse or a nutritionist. The role played by the doctor is enclosed within the concept of Telemedicine. It should be noted that the diabetic patient under a state of deep depression might stop his therapy based on Metformin, fact which leads to an excessive increase in glucose, and therefore the likelihood of an eventual risk to acquire diabetes irreversible complication is imminent. We do not rule out the possibility that the patient also stops his diet as a source of progressive glucose increase. Teleconsultation can be carried out only with the use of a cell phone with minimum requirements. In right panel of Fig. 3, the success probability falls down to 50% beyond 10 days fact that calls our attention in the sense that the Teleconsultation might not be so effective as one expects in a first instance. Clearly, having a poor telecentre infrastructure, and a small number of Teleconsultation, makes the impact weak and the number of patients with glucose above 200 remains constant over the next two weeks.

From a crude estimation having integrated Eq. 6. The success of the Tele-consultation demands at least 21 Teleconsultation per month, and that can be interpreted as

- 14 Teleconsultation on the endocrinologist side,
- and 7 on the psychological side.

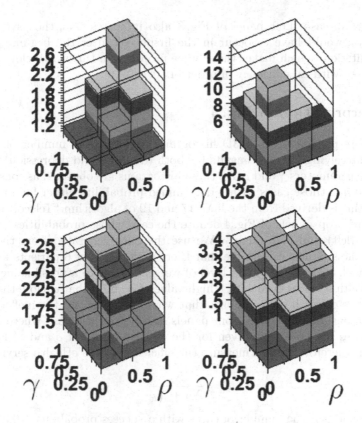

Fig. 4. Average number of patients versus the quantities ρ and γ up to for 4 scenarios: $P_F > 0.5$ (Top left), $P_F > 0.75$ (Top right), $P_S < 0.50$ (bottom left) and $P_S < 0.30$ (bottom right).

The effect of the 14 Teleconsultation per month is perceived as an attempt by the health professional to persuade patients to continue the prescription of Metformin daily. It is interesting to note that this scenario demands the usage of a 10 Mbps network that is accessible at low cost. The doctor can initialize a video call when have identified an anomalous event. We have considered the scenario where the efficiency is of the order of 0.5 as a realistic set-point. One can note that although cases with high glucose values prevail but with a greater number of patients with low number of depressive episodes. The simulations have shown that it is possible that the Teleconsultation scheme helps to counteract [8,9] the sudden increase of cases of depressive episodes. For a number of 60 patients with at least 5 depressive episodes, the proposed scheme indicates that at least 20 patients could reduce it by up to 1 per month, and with glucose values around 100 mg/dL. An important restriction in this study is that of the appearances of depressive symptoms are not exclusively caused by diabetes but could also have their origin in lateral complications derived from the apparition of an extra disease. Left panel of Fig. 3 also indicates that the psychological intervention would have to occur in the first week after the beginning of the Teleconsultation, which makes the scheme proposed in this work relevant, to be implemented with an efficiency of order of 50% [9,10].

6   Interpretation of Fig. 4

In Fig. 4 are plotted up to 4 3D histograms denoting the number of patients which are receiving the teleconsults for both (diabetes and depression) as function of the parameters γ and ρ. All these histograms are obtained as consequence of the counting those cases with a determined probability either fail or success. The counting is derived from the lines 17 and 19 of algorithm "Teleconsultations Simulation". Top panels of Fig. 4 denote the case whose probabilities of fail are above 0.5 (left) and 0.75 (right). We use the data generated from the Monte Carlo. In these panels we can see that there exists at least 6 patients for example left panel, with an imminent level of worsening due to a possible inefficiency of the eHealth services even with a high value of ρ and γ that enclose the performance of the eHealth services. Fact that would demonstrate the ineffectiveness of the teleconsults. In the bottom panels are plotted the case where P_S does not surpasses the 50% again even for the highest values of ρ and γ. For these histograms therefore we can measure the efficiency of the eHealth service as

$$\mathcal{E} = \frac{N_S}{N_S + N_F} \qquad (11)$$

where N_S and N_F the number of cases with a success probability >0.5 and fail probability >0.75 respectively. The resulting efficiency has turned out to be of order of 40%.

7   Conclusion

In this paper we have assessed in a quantitative manner the impact of a possible scenario of teleconsultations aimed to alleviate diabetes and depress progress in patients which are socially vulnerable.

According to our results, the probability of success for detaining the progress of both diabetes and depress might be below of the expected since it would depend strongly in resources in both available health specialists as well as a solid network that has null probability to loss connectivity so the phone and video calls are running permanently with external factors that can degrade the quality of the teleconsultation.

Despite of various limitations the success of teleconsultations to alleviate the pair: diabetes and depress might be interestingly high. Thus one expects that patients might recover in short times without the necessity of attending health centers and therefore to avoid translations fact that becomes the first factor to stop patients to attend to physical interviews with doctors [10] and psychologists due to their localization in peripheral zones of the city.

References

1. DeMaio, J., et al.: The application of telemedicine technology to a directly observed therapy program for tuberculosis: a pilot project. CID **33**, 2082–2084 (2001)
2. Palmer, A.J.: Computer modeling of diabetes and its complications. Diabetes Care **30**(6), 1638–1646 (2007)
3. Vistisien, D., et al.: Patterns of obesity development before the diagnosis of type 2 diabetes: the Whitehall II cohort study. PLoS Med. **11**(2), e1001602 (2014)
4. Nieto-Chaupis, H.: Prospects and expectations of ehealth services in north-Lima from mathematical modeling and computational simulation. http://ieeexplore.ieee.org/=7114467
5. Nieto-Chaupis, H.: Preventing risk situations at type-ii diabetes mellitus patients through continuous glucose monitoring and prediction-based teleconsults. http://ieeexplore.ieee.org/=7167450
6. Eli, K.: Diagnostic and Statistical Manual of Mental Disorders (DSM-5), 5th edn. American Psychiatric Association, Virginia (2013)
7. Numerical Recipes in C++, 2th version. Addisson-Wesley (1994)
8. Kamsu, B.: Systemic modeling in telemedicine. Eur. Res. Telemed. **3**(2), 57–65 (2014)
9. Nanda, P.: Quality of service in telemedicine. In: Digital Society. IEEE (2007)
10. Rispin, C.M.: Management of blood glucose in type 2 diabetes mellitus. Am. Fam. Phys. **79**(1), 29–36 (2009)

Optimal Dimensioning of Electrical Distribution Networks Considering Stochastic Load Demand and Voltage Levels

Esteban Inga[1]($\boxtimes$)(iD), Miguel Campaña[1]($\boxtimes$)(iD), Roberto Hincapié[2]($\boxtimes$)(iD), and Oswaldo Moscoso-Zea[3]($\boxtimes$)(iD)

[1] Universidad Politécnica Salesiana, Quito, Ecuador
{einga,mcampana}@ups.edu.ec
[2] Universidad Pontificia Bolivariana, Medellín, Colombia
roberto.hincapie@upb.edu.co
[3] Universidad Tecnológica Equinoccial, Quito, Ecuador
omoscoso@ute.edu.ec
https://www.ups.edu.ec/girei

Abstract. This work presents a model of optimal dimensioning of electrical distribution networks that uses real scenarios, georeferenced and contrasted by simulation processes that analyze the deployment and variables within the planning of electrical networks. This model considers a stochastic load demand and the voltage levels of the electrical distribution network. Moreover, this work exposes the sizing of the radial electrical network, the possible conditions to avoid a load imbalance and in this way, to prevent a system failure.

Keywords: Electrical distribution networks · Dimensioning
Optimization · Planning · Scalability · Smart grid

1 Introduction

This work warns of the need to optimally size electrical distribution networks of a model that allows scalability according to the users' demand, the efficiency, and reliability of the network. This is an essential requirement of the smart electrical network concept or Smart Grid (SG). For this, the work defines a problem of routing through graph theory that allows us to arrive at an optimal planning of the electrical network. This model defines a combinatorial problem that includes capacity constraints, costs, and profiles of the voltage of the network. Therefore, it is necessary to define a heuristic model that makes it possible to reach a near-optimal solution due to the characteristic of an NP-Complete model which lacks a globally optimal solution. These types of deployments: optimal and nearby of a network are commonly used in wireless communication networks or wire [5,13,14,16,26]. Reducing the costs for the used resources and ensuring

© Springer Nature Switzerland AG 2018
A. D. Orjuela-Cañón et al. (Eds.): ColCACI 2018, CCIS 833, pp. 200–215, 2018.
https://doi.org/10.1007/978-3-030-03023-0_17

the connectivity of its users. In addition to providing for the technology to be deployed [15, 25]. Analogously, this work proposes a tree of minimum expansion to achieve the connectivity of users within a georeferenced area and automatically engage electrical distribution lines in a tree or radial topology configuration contemplating a medium voltage circuit and other with a low voltage circuit for connectivity to the users.

The model is developed considering a multigraph that defines a layer for the medium voltage network and another layer for the low voltage network, ensuring the connectivity of users in the deployment depending on demand. To achieve the objectives a deployment is made in urban areas using an osm file with the information of longitude and latitude within a model developed in Matlab and simulated in Cymdist [24].

Medium and low voltage networks have been treated as separate problems [1, 9]. The model is distinguished from other works, in four aspects of great importance: (a) it uses and manages the georeferenced information obtained from a .osm file provided by a free platform; which, contains detailed information of locations, surfaces, characterization of zones by their type (residential, commercial or educational) as presented in [2], (b) unlike other methods described by [10, 22] an inductive method is applied in this research. This method starts from the topographic reality and consumption levels for each subscriber until achieving the route map and dimensioning of the distribution network equipment. This guarantees the minimum resources to be used to satisfy the service to final consumers observing levels of voltage drop, both in medium voltage and low voltage an in this way achieving the best topology, (c) costs were reduced by using grouping techniques [4] designed to achieve the objective in this research, leaving aside traditional clustering techniques such as k-means, k-medoids or mean shift that restrict the search space, discarding the possibility of finding the best solution to the planning problem [21]. In this work, trees of minimum expansion (medium voltage network) and maximum expansion (low voltage network) are used and through the setcover algorithm a solution to the problem of location of the transformers by selecting the best subsets is given, and (d) the present model admits restrictions and real capacity parameters, maximum ranges of linear coverage in the secondary network, demands for each type of user, sectorization according to the strata, classification by the levels of consumption. In this context a close to the optimum route map can be build in a controlled way which uses real parameters of great importance in the design stage.

Henceforth, this article is organized as follows. Section 2 presents the relationship between theory of graphs for an optimal dimensioning and planning of an electrical network. The formulation of the problem is described in Sect. 3. Section 4 analyzes the results of the model and its simulation. Finally, we conclude this article in Sect. 5.

2   Dimensioning of Electrical Distribution Networks

2.1   Planning of Electrical Distribution Networks

In order to relate sizing to achieve a near and optimal deployment of an electrical distribution network, there are certain needs that are commonly discarded in the planning of a distribution system and are linked to the scalability, the minimum cost for employed resources and those that are considered in a network area or in any of their topologies. The costs involved in planning processes are: the conductor, civil works, transformers, among others. On the other hand, the planning of the network commonly does not consider a growth of loads in consideration of technological progress or by the inclusion of new basic items such as an electric car [12] or an induction cooker. The network does not have mechanisms to transform itself from a basic distribution network in a resilient network [3] and being able to self-regulate load or to get a new driver's route for realizing a process of enlargement or reconfiguration by emerging situations caused by natural phenomena.

The continued increase of buildings requires an increase of underground networks and a decrease of air networks, including the visual aspect that requires formalizing a suitable atmosphere of a modern city and less prone to damage from natural disasters. A network of underground distribution must contemplate a civil work that includes pipelines, wells, cameras, concrete base for transformation, electrical work, equipment, accessories, cables and conductors. In addition, the design must consider the security of the pipeline with a minimum depth according to rules, the conductor with a certain isolation, the voltage level, the nominal current, the short circuit current, the thermal, mechanical and environmental conditions, and the lifetime. For this reason, a electrical distribution network requires a reduced risk of investment. For this work a medium voltage network (MV) and a low voltage network (LV) in a hub-and-spoke topology with an unidirectional flow in the recipient direction are considered. The model for a residential load will be able to receive a service of 0.22/0.120 kV of a local distribution network, the same that has a behavior of hours valleys in function of the high demand in the early hours of the morning and evening. In this way, the network will have a sensitive load and will be linked to the ambient temperature. The model presents a network dimensioned in two layers or subsystems: (a) primary network (MV) (b) secondary network (LV), where the primary network will be responsible for transporting the energy between the substation and distribution transformers. While the secondary network will handle the final link with the customers and is determined by their supply. For this deployment we have a primary network originated in the substations that may have a aerial or underground transition forming a minimum expansion tree or medium voltage trunk with a derivation to the transformation chambers in a range of tension levels between 6.8 kV and 13.2 kV. In urban areas, the primary feeders form a radial circuit that may be in buried pipelines and the secondary networks must be properly sized to avoid failures or imbalances that would decrease the reliability of the network or the quality of the power supply [3].

In an underground network an operation should be considered in a hub-and-spoke topology with an open connection to the closest feeders. A lateral circuit of the type single-phase or three-phase with nominal current will also be in open mode. Switches or isolated cable connectors can be considered. The protection system of an underground network is determined in connection with the site of the failure in two stages, (a) primary cable failure with a reset operated by the circuit switch of the feeder at the substation and (b) the fuses cut the failure in the lateral circuit.

2.2   Minimum Spanning Tree Under Optimality Criteria

When considering a model based on graph theory we can associate an optimal routing of an electrical network that covers the minimum spanning tree at the lowest cost, for which, different options have been proposed based on the Prim or Kruskal algorithms for realizing the routing of the distribution transformers and thus, achieving the medium voltage network. For this reason, it should be taken into account the loads-users georeferenced coordinates and the units that host these loads. Thus, it is necessary to consider the intersections of the streets and avenues as candidate sites to locate a distribution transformer. But, depending on the load and capacity, the number of active transformers are minimized in the deployment. Aditionally, the connectivity of users in a low voltage network is guaranteed. The model starts from a minimum spanning tree [19,23] and other complements to reach a minimum Steiner tree [11,18]. Other works propose a previous segmentation of the area for the network deployment through triangulation with Delaunay and Voronoi [6,17,20,28], But in this work, it will not be considered.

A problem of dimensioning, for the deployment and planning of a distribution electrical network that takes into account the demand as a non-trivial issue, requires a result close to the optimal allowing electric distribution companies to reduce the design times in the planning phase by ensuring that the cost of the investment is minimal [8], Thereby, generating efficiency in the network and by using scalable models of planning allows us to optimally predict future demand and meet new requirements while maintaining the reliability and energy security [7,27]. A process of additional simulation and parallel to the optimization model gives us an analysis of the power flow, balance sheet or off-balance load, failures, estimation of loads that may have an impact on the model of network routing. Within the sizing and planning of a distribution network, variables such as the topology and its maximum permitted levels of voltage drop must be considered. It is also important to consider as well the maximum load of a substation in accordance with their capabilities that correspond to the chargeability within a accessible limit. There are other variables that are not considered in the present work as restrictions of lines and conductors but will be considered in future work. It is important to mention that, for the calculation of the point-to-point lengths of each section, the model will use the Manhattan and Haversine distance. In the first case, contributes in the model to avoid diagonals as possible paths in the graph. The Harversine distance while contemplating the spherical shape of the earth, provides us with more accurate data of lengths in km [13]. For the simulation process it is considered a maximum demand by user type (residential, commercial, or industrial), transformers type, nominal voltage for the primary and secondary voltage limits and the type of conductor considering that the network may include an underground or aereal deployment.

3   Problem Formulation

To solve the combinatorial problem defined as NP-Complete, proposed in this work, a heuristic model is included. This model supports to find the minimum cost to achieve the maximum percentage of connectivity to the distribution network of the final consumers.

In the following Eq. 1 the total cost on each transformer of each cluster is presented, where ze is the number of elements of each cluster and Cp is the power consumed by each user. This was tested with variable demands according to the degree of consumption. In the Eq. 2, C_{sub} is the total cost of the substation to be deployed in the area of interest, le is the number of existing clusters in the scenario.

$$Ctr = \sum_{k1=1}^{ze} Cp_{(k1)} \tag{1}$$

$$C_{sub} = \sum_{k1=1}^{ze} \sum_{p=1}^{le} Ctr_{(k1,p)} \tag{2}$$

In the Eq. 3, the total cost of the medium voltage network driver is represented by Cmv_{con}, the expression d_a is the length of the section from point i to j, the multiplication factor k that will depend on the conductor section to determine its cost and finally N is the length of the vector that contains the positions of the transformers including the position of the substation. The expression $N - 1$ represents the minimum number of edges to guarantee connectivity to all the distribution transformers from the substation.

$$Cmv_{con} = \sum_{a=1}^{N-1} d_a * k \tag{3}$$

In the Eq. 4, the total cost of the low-voltage network conductor is presented. The expression $d_{(h,a)}$ represents the maximum distances required in each section of each cluster to guarantee connectivity to the end users from their respective transformers.

$$Cbv_{con} = \sum_{h=1}^{le} max \sum_{a=1}^{zise(ze)-1} d_{(h,a)} * k \tag{4}$$

Therefore, the optimization problem is defined as follows:

$$min\ Ctr + C_{sub} + Cmv_{con} + Cbv_{con} + \sum_{k=1}^{ze} Cp_{(k)} \tag{5}$$

Subject to:

$$\sum_{s,k\ \in\ n} (s - 1) = n, \ \forall\ s, k \in n; \forall\ n \in A(n) \tag{6}$$

$$\sum_{s \in S} S \leq m, \ \forall\ S \in A(n); \forall\ m > 1 \tag{7}$$

$$X = \sum_{r_{i,j} \in r_{ds}} r_{tm} \leq d_{max}, \ \forall\ X \in A(n) \tag{8}$$

The Eq. 5 corresponds to the objective function, which consists of minimizing the costs of implementation in an electrical network of buried distribution. The Eq. 6 restricts the algorithm to form a minimum expansion tree from the substation to the primary ones of the electrical distribution transformers. The topology in graph theory tree type in electrical distribution networks is known as radial topology.

The capacity restriction, of the Eq. 7, limits the number of users that will be able to agglutinate each transformer and later determine the required capacity in kVAs. In such a way that the distribution transformer will be able to supply electric power to all the demand.

The Eq. 8 restricts the maximum distance allowed for an end user to be part of a distribution transformer.

Algorithm 1 OSEDN - Optimal Sizing of Electrical Distribution Networks

Step: 1 Variables

$m, r_{tm}, Lon, Lat, S_x, S_x, X_{int}, Y_{int}, n, dist, d_{traf}$ use, $flag, k, z, l, x_t, y_t, p, G, G1,$

Step: 2 Initialization

$[X_{int}\ Y_{int}]\ \leftarrow\ connectivity\ matrix$

X= $[Lat\ Lon]$; n= $length[lat]$

Step: 3 Calculating

$dist_{i,j}\ \leftarrow\ haversine[Lat\ Lon]$

Step: 4 Clustering

while use $\leftarrow$ 1 **do**

$flag = 1$

while $flag\ \leq\ m$ **do**

for $i\ \rightarrow\ length(n)$

$k\ \rightarrow\ (min(min(dist)))$

if $length$ k $\neq$ 0

[row col]= $find(dist == min(min(dist)))$

$use = [use\ X(i)]$

$next$

elseif

$flag == 0$

endif

if $length$ use> m

$flag\ \leftarrow\ 0$

endif

endfor

endwhile

endwhile

$tmp \leftarrow use$

Step: 5 Connectivity matrix

while $z\ \leq\ n$ **do**

$G(tmp_{ij}, tmp_{ji})=1$

z=z+length(tmp)

Step 4 $\leftarrow$ $return$

endwhile

Step: 6 Transformer selection

$x_t\ \leftarrow\ sum(Lon)/length(Lon)$

$y_t\ \leftarrow\ sum(Lat)/length(Lat)$

p $\leftarrow$ $[x_t, y_t]$

$l\ \leftarrow\ find\ position$ p

Step: 7 MST

$d_{traf}\ \leftarrow\ Manhattan\ [Lon\ Lat]$

$G1(d_{traf}\ \leq\ dmin) = 1$

path $\leftarrow$ $prim_{mst}(sparse(G1))$

---

4   Results Analysis

In this section, the operation and potential of the proposed heuristic proposed in the planning and deployment of electrical distribution networks will be validated. Two important moments of analysis will be considered. In the first moment, through the Matlab software, a process of resource optimization will be carried out. In addition, the necessary information of the elements to be located in the geo-referenced area will be obtained, such as: installed demand and demand consumed by each transformer, length and minimum route of the trench required and sizing of the substation. Therefore, through the Matlab software we obtain the route map of the electric network to be deployed observing geographical positions of the distribution equipment. Also considering technical parameters in distribution networks, such as: voltage drops in medium and low voltage in order to guarantee a quality and highly reliable service. In a second moment of analysis, using Cymdist software, a power flow will be run and the behavior of electrical parameters will be observed, such as: voltage drops in the medium voltage network, current flowing through the conductor in each section from the substation to each of the primary transformers, the consumed demand (kVA) in each transformer and the reactive power that crosses in the conductors from the source to the service nodes. Next, in Table 1, we present the simulation parameters. This table describes the general characteristics to be considered as initial criteria for the planning and deployment of underground electrical networks.

Table 1. Model simulation and parameters

Deployment	Users density	2057 Users/km^2
	Area	0.2217 km^2
	Users location	Georeferenced
	Capacity per cluster	25 users
	Coverage range	0.02 km^2
Aplication	Residential demand c/a	Varying per stratum kVA
	Concentrated load	Unbalanced
	Installed capacity	4.5 MVA
	Voltage MV/LV	13.2 kV/0.22 kV

The optimal deployment and planning of underground distribution electric networks is presented in Fig. 1. In this figure the georeferenced location of the distribution transformers, the location of wells, the substation, the user location and the route through which the power lines of the primary and secondary network must pass are shown. Therefore, the proposed model is able to provide all the necessary data to build a new electrical network for buried distribution.

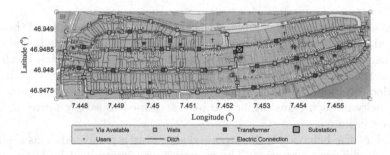

Fig. 1. Planning of electrical distribution networks - Source: Author

Figure 2 shows the voltage drop in function of the distance in the secondary circuits of each distribution transformer located in the georeferenced area. With the aforementioned figure, it can be seen that none of the transformers in their different tested scenarios exceeds the levels of voltage drops provided in the literature from the source to the furthest user.

In addition, Fig. 2 shows that as the ability to group users increases, the service distance from the source to its most distant node also increases. This happens because if the capacity to agglutinate users increases, the capacity in kVAs required by the distribution transformer to satisfy the demand of that group of users also increases. Consequently the distance of linear coverage increases. This is mainly because the transformer is more robust and can assume this load. Another interesting fact is that as the capacity increases the number of distribution transformer units required in the deployment decreases. This will cause the installed power required in each distribution transformer to increase. This will be explained in detail in the following paragraphs.

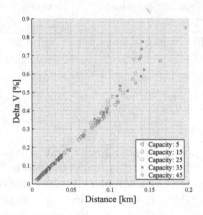

Fig. 2. Voltage drop as a function of the distance in the secondaries of the distribution transformers - Source: Author

The value of installed capacity in the distribution transformers and the power consumed by the users at full load are illustrated in Fig. 3. Through this metric people can have an overview of the state of chargeability in each of the distribution transformers. In addition, the IDs of the transformers that may be overloaded can be identified. Therefore, with this metric the percentage of available capacity in each transformer can be determined to satisfy future increases in power consumed.

In Table 2 the behavior of the proposed algorithm is presented. This algorithm is tested in different scenarios. The scenarios are determined by the variation of the grouping capacity for the conformation of the conglomerates, which is illustrated in column 1 of this table. Consequently, it is appreciated that, as the ability to host users in different clusters, the algorithm delivers different percentages of coverage being the maximum value obtained in scenario 3, with a capacity of 25 users (see Table 2), reaching 72.37%, which means that 126 users were not associated to a cluster.

The reason why some users did not manage to be part of a conglomerate is explained by the restrictions of maximum coverage range and grouping capacity. This restrictions depending on the set point serve to restrict the users' membership of those who exceed the initial conditions of the problem. Therefore, those users who were not covered, either because they did not reach the maximum range of coverage or because the surrounding conglomerates were already at maximum capacity will be subject to analysis at the design stage of the distribution network.

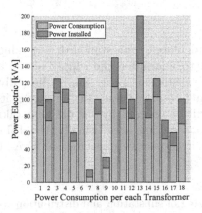

Fig. 3. Amount of reserve in distribution transformers at full load distributed in the area of interest - Source: Author

The option that may arise as alternative is the location of extra transformers to those deployed by the heuristic or to place transformers of greater capacity than those suggested by the model and allow coverage to the user that requires it until covering with 100% of electric service. In addition, in Table 2, it can be observed that when grouping 456 users in different scenarios, the cpu time that the computer uses is directly proportional to the capacity. This means that as the number of maximum elements that belong to a cluster increases the time in which the computer provides the solution to the planning problem increases. This happens because the number of combinations for the best solution search increases. Then again, if the capacity to host users in a conglomerate increases, the number of transformers to be emplaced decreases. Finally, if the number of transformers to be placed decreases, it must necessarily increase the installed capacity in the distribution transformers, as shown in Table 2.

Table 2. Planning of electrical distribution networks tested in different scenarios with 456 users - Source: Author

Capacity	Coverage (%)	Time (sec)	# Needed transformers
5	58.33	73.15	58
15	64.42	124.41	25
25	72.37	180.10	18
35	59.21	231.21	12
45	68.86	331.10	11

Table 3 shows the calculated (non-standardized) value in MVAs of the substation design required to guarantee the supply of the demanded power. The calculated value of the substation is the sum of the individual powers required considering an additional reserve value. In the model it is entered a variable in percentage, which will be entered by the designer taking into account studies of demand projections in that area. Finally, the Table 3 shows the number of transformers and their respective capacities in kVAs to be displayed in the georeferenced area.

Table 3. Sizing of substation and distribution transformers.

Capacity	Sizing substation (MVA)	Power instaled of transformers (kVA/Units)
5	1.84	30/35; 45/1; 15/20
15	1.85	75/5; 45/3; 60/4; 50/3 100/7; 15/3
25	1.86	112.5/2; 100/5; 125/4; 60/1; 15/1; 30/1; 150/1; 200/1; 75/1; 50/1
35	1.85	125/1; 150/2; 200/3; 100/3; 75/1; 45/1; 15/1
45	1.88	250/2; 200/2; 75/1; 150/1; 100/3; 225/1; 15/1

Figure 4 shows the average power consumed per meter in the substitutes of the distribution transformers in each scenario tested. It can be seen that as the ability to group the users increases the apparent power that will circulate through the driver from the source (distribution transformer) to the furthest service node will also increase. The metrics obtained in Fig. 4 help to size the optimal section of the conductor required for the transport of the power demanded by the end users. Therefore, as a conglomerate increases in number of users, there is a cumulative need to increase the installed power of the distribution transformer, the gauges of the conductor and the electrical protections for the different circuits (primary or secondary). Therefore, the operation of the proposed algorithm in the present document for the design and planning of electricity distribution networks is demonstrated. This provides a road map that serves as an initial guide to the electrical designer of distribution networks and for later analysis with specialized software such as Cymdist.

From here on we will make an analysis on the route map obtained by the heuristic in order to determine the feasibility of implementation of the proposed solution.

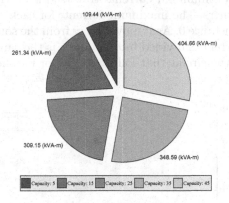

Fig. 4. Average apparent power that would pass through a conductor for each meter - Source: Author

4.1   Analysis Using Cymdist Software

In this section, the route map and the electric distribution network obtained in the previous section will be modeled. To do this, the Cymdist software will be used, which will allow us to simulate a medium voltage power flow. Consequently, Fig. 5 presents the topology obtained by the proposed heuristic with the modeling of the medium voltage network. A three-phase unbalanced network will be modeled with the consumption data of each transformer obtained in the previous section.

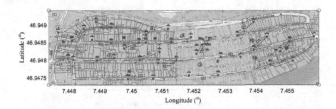

Fig. 5. Power flow in the medium voltage network - Source: Author

The voltage drops, from the substation to each of the primary transformers, are represented in Fig. 6. As the distance from the substation increases the voltage levels until the primary decreases. Consequently, the distance and characteristics intrinsic to the conductor (resistivity) means that there is more or less heat dissipation in each section. This heat dissipation is known as the Joule effect. In addition, it can be seen that the maximum voltage levels allowed (in percentage) described in the literature do not exceed 5% from the substation to each of the primary distribution transformers. With Fig. 7 it is possible to identify the maximum and minimum currents traversed from the source to each of the primary transformers. The maximum current, for each phase, is reflected in the substation with distance 0. As it moves away from the source, with its respective topology, the current is divided into nodes or is the same. This depends on the route or path of the circuit that follows (serial or parallel).

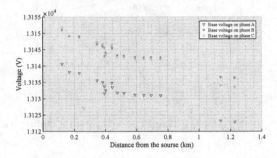

Fig. 6. Voltage drop as a function of the distance in the primary of the distribution transformers - Source: Author

Therefore, in Fig. 7, after knowing the maximum current that will flow through each section and the lengths (in km), it will be possible to dimension the conductor, the protections and the necessary equipment for the implementation of an optimal system of electrical distribution networks.

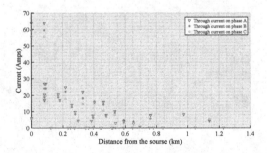

Fig. 7. Current flowing through the conductor, for each phase, depending on the distance - Source: Author

Therefore, it is demonstrated that, through an optimization process applying graph theory it is possible to construct an optimal road map that allows us to design and plan electricity distribution networks. This route map obtained by the proposed heuristic is the starting point for a reliable design process. Since the model admits income variables, such as: capacity and coverage. Moreover, through a load vector allows us to enter representative loads to each user. Which is identified according to their stratum by observing the amount of consumption.

5   Conclusions

In the present work an algorithm capable of planning robust electrical distribution networks is proposed. This enables the possibility of integrating optical fiber to provide intelligence to the electricity distribution network through the implementation of robust communication systems, which will allow us to monitor the current state of the network at all times. The proposed Heuristic considers real electrical parameters, such as, voltage drops in the medium and low voltage network, chargeability in the transformers, demands (kVA) of different strata, section of the conductor to be used in medium and low voltage networks and capacity of distribution transformers.

Therefore, the fundamental contribution of the present research is to provide an initial route map of the topology to be deployed in an electrical distribution network expanding the possibility of integrating Fiber Optics for communication networks. Moreover, considering real scenarios that allow the designer in the initial stage of design to make decisions based on the results obtained from the optimization heuristic that conjugated with specialized simulation software for distribution networks such as Cymdist can contribute in a significant way in the stages prior to the construction of the network. This guarantees that the final results are optimal at the lowest cost without neglecting the technical-economic criteria, energy security, quality, reliability and continuity of service.

Acknowledgment. This work has been produced thanks to the support of the GIREI - Intelligent Electrical Networks Research Group of the Universidad Politécnica Salesiana Ecuador under the project Infrastructure of Advanced Measurement and Response of Electric Power Demand in Smart Grid.

References

1. Abeysinghe, S., Wu, J., Sooriyabandara, M., Abeysekera, M., Xu, T., Wang, C.: Topological properties of medium voltage electricity distribution networks. Appl. Energy (2017). https://doi.org/10.1016/j.apenergy.2017.06.113
2. Alhamwi, A.: OpenStreetMap data in modelling the urban energy infrastructure: a first assessment and analysis. Energy Procedia **142**, 1968–1976 (2017). https://doi.org/10.1016/j.egypro.2017.12.397
3. Anuranj, N.J., Mathew, R.K., Ashok, S., Kumaravel, S.: Resiliency based power restoration in distribution systems using microgrids. In: 2016 IEEE 6th International Conference on Power Systems (ICPS) (2016). https://doi.org/10.1109/ICPES.2016.7584186

4. Bajpai, P., Chanda, S., Srivastava, A.K.: A novel metric to quantify and enable resilient distribution system using graph theory and choquet integral. IEEE Trans. Smart Grid **3053**(c), 1 (2016). https://doi.org/10.1109/TSG.2016.2623818
5. Campaña, M., Inga, E., Hincapié, R.: Optimal placement of universal data aggregation points for smart electric metering based on hybrid wireless. In: CEUR Workshop Proceedings, vol. 1950 (2017)
6. Chen, Y.: Optimal weighted Voronoi diagram method of distribution network planning considering city planning coordination factors. In: 2017 4th International Conference on Systems and Informatics (ICSAI) (Icsai), vol. 1, pp. 335–340 (2017). https://doi.org/10.1109/ICSAI.2017.8248314
7. Davidescu, G., Stutzle, T., Vyatkin, V.: Network planning in smart grids via a local search heuristic for spanning forest problems. In: IEEE International Symposium on Industrial Electronics, pp. 1212–1218 (2017). https://doi.org/10.1109/ISIE.2017.8001418
8. Dimitrijevic, S., Rajakovic, N.: Service restoration of distribution networks considering switching operation costs and actual status of the switching equipment. IEEE Trans. Smart Grid **6**(3), 1227–1232 (2015). https://doi.org/10.1109/TSG.2014.2385309
9. Esmaeeli, M., Kazemi, A., Shayanfar, H.A., Haghifam, M.R.: Sizing and placement of distribution substations considering optimal loading of transformers. Int. Trans. Electr. Energy Syst. Int. (2014). https://doi.org/10.1002/etep
10. Gouin, V., Member, S., Raison, B.: Innovative planning method for the construction of electrical distribution network master plans. Sustain. Energy Grids Netw. (2017). https://doi.org/10.1016/j.segan.2017.03.004
11. Han, X., Liu, J., Liu, D., Liao, Q., Hu, J., Yang, Y.: Distribution network planning study with distributed generation based on Steiner tree model. In: 2014 IEEE PES Asia-Pacific Power and Energy Engineering Conference (APPEEC), vol. 1, pp. 1–5 (2014). https://doi.org/10.1109/APPEEC.2014.7066185
12. Hemphill, M., South, N.: Electricity distribution system planning for an increasing penetration of plug-in electric vehicles in New South Wales. In: 2012 22nd Australasian Universities Power Engineering Conference (AUPEC), pp. 1–6 (2012)
13. Inga, E., Céspedes, S., Hincapié, R., Cárdenas, A.: Scalable route map for advanced metering infrastructure based on optimal routing of wireless heterogeneous networks. IEEE Wirel. Commun. **24**(April), 1–8 (2017). https://doi.org/10.1109/MWC.2017.1600255
14. Inga, J., Inga, E., Hincapié, R., Cristina, G.: Optimal planning for deployment of FiWi networks based on hybrid heuristic process. Latin Am. Trans. IEEE (Revista IEEE Am. Latina) **15**(9), 1684–1690 (2017). https://doi.org/10.1109/TLA.2017.8015053
15. Inga-Ortega, E., Peralta-Sevilla, A., Hincapié, R.C., Amaya, F., Tafur Monroy, I.: Optimal dimensioning of FiWi networks over advanced metering infrastructure for the smart grid. In: 2015 IEEE PES Innovative Smart Grid Technologies Latin America (ISGT LATAM), pp. 30–35 (2015). https://doi.org/10.1109/ISGT-LA.2015.7381125
16. Inga-ortega, J., Inga-ortega, E., Gómez, C.: Electrical load curve reconstruction required for demand response using compressed sensing techniques. In: IEEE PES Innovative Smart Grid Technologies Conference - Latin America (ISGT Latin America) (2017). https://doi.org/10.1109/ISGT-LA.2017.8126731
17. Jiménez-Estévez, G.A., Vargas, L.S., Marianov, V.: Determination of feeder areas for the design of large distribution networks. IEEE Trans. Power Deliv. **25**(3), 1912–1922 (2010). https://doi.org/10.1109/TPWRD.2010.2042468

18. Kahveci, O., Overbye, T.J., Putnam, N.H., Soylemezoglu, A.: Optimization framework for topology design challenges in tactical smart microgrid planning. In: 2016 IEEE Power and Energy Conference at Illinois, PECI 2016, pp. 1–7 (2016). https://doi.org/10.1109/PECI.2016.7459262

19. Li, J., Ma, X.Y., Liu, C.C., Schneider, K.P.: Distribution system restoration with microgrids using spanning tree search. IEEE Trans. Power Syst. **29**(6), 3021–3029 (2014). https://doi.org/10.1109/TPWRS.2014.2312424

20. Lu, Z., Wang, S., Ge, S., Wang, C.: Substation planning method based on the weighted Voronoi diagram using an intelligent optimisation algorithm. IET Gener. Transm. Distrib. **8**(October 2013), 2173–2182 (2014). https://doi.org/10.1049/iet-gtd.2013.0614

21. Mam, M.: Improved K-means clustering based distribution planning on a geographical network. I. J. Intell. Syst. Appl. **9**(April), 69–75 (2017). https://doi.org/10.5815/ijisa.2017.04.08

22. Mateo, C., et al.: Electrical power and energy systems European representative electricity distribution networks. Electr. Power Energy Syst. **99**(July 2017), 273–280 (2018). https://doi.org/10.1016/j.ijepes.2018.01.027

23. Montoya, D.P., Ramirez, J.M.: A minimal spanning tree algorithm for distribution networks configuration. In: IEEE Power and Energy Society General Meeting, pp. 1–7 (2012). https://doi.org/10.1109/PESGM.2012.6344718

24. Nagarajan, A., Ayyanar, R.: Application of minimum spanning tree algorithm for network reduction of distribution systems. In: North American Power Symposium (NAPS), pp. 1–5 (2014). https://doi.org/10.1109/NAPS.2014.6965353

25. Peralta, A., Inga, E., Hincapié, R.: FiWi network planning for smart metering based on multistage stochastic programming. Latin Am. Trans. IEEE (Revista IEEE Am. Latina) **13**(12), 3838–3843 (2015). https://doi.org/10.1109/TLA.2015.7404917

26. Peralta, A., Inga, E., Hincapié, R.: Optimal scalability of FiWi networks based on multistage stochastic programming and policies. J. Opt. Commun. Netw. **9**(12), 1172 (2017). https://doi.org/10.1364/JOCN.9.001172

27. Shi, J., Qiao, Y., Wang, Y., Wen, J., Tong, J., Zhang, J.: The planning of distribution network containing distributed generators based on mixed integer linear programming. In: 2015 5th International Conference on Electric Utility Deregulation and Restructuring and Power Technologies (DRPT), pp. 5–9 (2015). https://doi.org/10.1109/DRPT.2015.7432303

28. Wang, S., Lu, Z., Ge, S., Wang, C.: An improved substation locating and sizing method based on the weighted Voronoi diagram and the transportation model. J. Appl. Math. **24**, 1–8 (2014). https://doi.org/10.1155/2014/810607

Author Index

Printed in the United States
by Bookmasters

Printed in the United States
By Bookmasters